STUDY GUIDE A

Leslie Kinsland
University of Southwest Louisiana

Chemistry and Our Changing World THIRD EDITION

Alan Sherman
Middlesex Community College

Sharon J. Sherman
Rutgers University

PRENTICE HALL, *Englewood Cliffs, New Jersey 07632*

Editorial/Production Supervision: *Sandra Lynn Barrett*
Supplements Acquisitions Editor: *Alison Muñoz*
Prepress Buyer: *Paula Massenaro*
Manufacturing Buyer: *Lori Bulwin*

© 1992 by Prentice-Hall, Inc.
A Simon & Schuster Company
Englewood Cliffs, New Jersey 07632

Printed in the United States of America

10 9 8 7 6 5 4 3 2 1

ISBN 0-13-131582-X

Prentice-Hall International (UK) Limited, *London*
Prentice-Hall of Australia Pty. Limited, *Sydney*
Prentice-Hall Canada Inc., *Toronto*
Prentice-Hall Hispanoamericana, S.A., *Mexico*
Prentice-Hall of India Private Limited, *New Delhi*
Prentice-Hall of Japan, Inc., *Tokyo*
Simon & Schuster Asia Pte. Ltd., *Singapore*
Editora Prentice-Hall do Brasil, Ltda., *Rio de Janeiro*

CONTENTS

iii

TO THE STUDENT

This study guide is intended as a supplement to <u>Chemistry and Our Changing World</u> 3/e by Alan Sherman and Sharon J. Sherman. It is designed to aid in the understanding and the application of material introduced in the main text. It also provides a means of monitoring your comprehension of the various topics and a way of checking your preparation for examinations in the course. The study guide should be used as a complement to rather than as a substitute for the main text.

This book is arranged in the following manner:

Chapter Overview: This gives a summary of the main concepts discussed in the chapter. It focuses your attention upon the broad picture rather than upon specific points. It should remind you of the overall emphasis of the material that was presented.

Topic Summaries and Examples: This provides a section by section review which is labeled by the same specific headings as those within the text. Key words are highlighted. Examples with explained solutions are provided to illustrate topics which often prove more difficult for students.

Self-Test: This is a collection of test questions of the sort which are often used by instructors. The self-test checks your comprehension and mastery of the material within the chapter. It, therefore, provides a means of monitoring your preparation for actual course exams.

Answers to the Self-Test: The answers to the self-test are provided. Elaboration on why a particular answer was chosen is presented when the choice may not be obvious from rereading either the text or the topic summary.

To maximize the benefits that you will derive from this book and from the course you should follow these study procedures:

1. Consult the authors' chapter objectives and read the main text before going to class. This will give you a general idea about what the instructor may cover.

2. Attend class and carefully note the topics that the instructor stresses. List the problems and objectives which the instuctor assigns from the text.

3. Reread the main text and work the problems assigned. Review any areas of difficulty.

4. Read the chapter overview and the topic summaries with their examples in this study guide.

5. Work the self-test and check your answers.

6. Consult this study guide and the main text to clarify the answers for any questions that you missed.

7. Review all of your previous work, stressing the areas which have given you problems, before each exam in the course.

As you go through this course, remember that chemistry is all around you and that learning about it should be an adventure rather than a chore. Watch for chemical references in magazines, in newspapers and on television. Look at the ingredients listed for your foods, soaps, cosmetics. Correlate your observations of nature and your surroundings to what you learn in the course. Use this information in your everyday life, rather than putting the knowledge and text book out of sight and mind. Let the information be useful and beneficial in the **real** world as well as in satisfying some university science requirement.

I would like to thank Dan Joraanstad for involving me in the reviews of the Shermans' text and Thomas Henry Moore for suggesting that I prepare this study guide. I also want to acknowledge the invaluable proofreading, comments and suggestions provided by Dr. Richard S. Perkins, Professor of Chemistry at the University of Southwestern Louisiana. My daughter Cynthia L. Kinsland, herself a recent chemistry student and thus closer to the "receiving end" of this information, not only proofread the manuscript, but requested modifications which she felt would be beneficial to other students undertaking the study of this subject.

1

SCIENCE, TECHNOLOGY, HUMANKIND, AND THE ENVIRONMENT: AN INTRODUCTORY CHAPTER

CHAPTER OVERVIEW

This chapter introduces sciences in general and then focuses on chemistry in particular. In doing so, it lays the groundwork upon which the rest of the course is built. The scientific method and the means by which science develops are discussed. The relationship and the differences between science and technology are emphasized. Some varied current applications of chemistry are presented. Emphasis is placed upon both the benefits and risks of technological advances. A rationale for studying chemistry is presented in that a scientifically informed populace can make not only personal decisions but can also aid in community decisions about issues which deal with scientific matters.

TOPIC SUMMARIES AND EXAMPLES

Introduction

Chemistry is described as being the central science upon which our entire lives depend. Examples such as the fact that our bodies and the foods that fuel them are complex chemical systems are presented to justify that statement. A scenario describes the type of reasoning which leads to possible chemical solutions to medical problems. In it, the drug t-PA is mentioned as being a possible useful treatment for stroke patients because of its success in dissolving clots and because of its performance in animal cases.

Science and Technology

Science is the systematic study of the world around us. The **scientific method** of approaching the world involves: (1) observation and description; (2) searching for patterns;

(3) making generalizations and formulating laws, which are simple statements or equations describing a natural phenomenon; (4) proposing **hypotheses**, which are educated guesses to explain observed laws, and developing **theories**, which are widely accepted explanations; (5) making predictions based on the theories; and, (6) testing theories by experimentation, changing one variable at a time. Theories may be revised or discarded completely on the basis of new data acquired. The phlogiston theory of combustion is discussed as an example of a theory which did not stand the test of further experimentation and was later replaced.

The scientific method may be applied to events and objects in our general lives as well as to what we consider specifically scientific matters.

Example 1.1

A woman tries to start her car; it does not start (an observation). She also notices that the headlight switch is on (an observation). She knows that leaving the lights on will drain the battery (a law). She believes that the car did not start because the battery was dead (a hypothesis or a theory). She presumes that the car will start if the battery is recharged (making predictions based on a theory). She has the battery recharged and the car indeed starts (testing the theory, one variable at a time). If the car had still not started, additional observations would have been required, more laws consulted and further theories proposed and tested.

Chemistry is the study of matter and the changes it undergoes. **Matter** has mass, occupies space and can be detected by our senses. **Mass** is the measure of the amount of matter that an object contains. **Weight** is distinguished from matter in that weight depends upon both the mass contained in an object and the force of gravity to which the object is subjected. In this course, mass is to be studied both qualitatively and quantitatively. In **qualitative** work, properties of matter are discussed, while in **quantitative** work, meaningful numbers are applied to some aspect of matter.

Chemistry is one of the **natural sciences** which include physics, geology, astronomy and the biological sciences as well as all of their hybrids such as geochemistry and astrophysics. These are studies of natural phenomena and are thus diffferent from the social sciences such as anthropology, psychology and sociology which are involved with the vagaries of humankind. **Basic science** involves the obtainment of knowledge for its own sake. **Applied science** is the attempt to acquire knowledge to be used for solving a particular problem. **Technology** is the application of science to produce

2

material goods.

Example 1.2

Basic science gives us the information that a wire will glow if electricity is passed through it. Thomas Edison utilized that knowledge and worked toward applying it as a light source. The technology which he developed is known as the incandescent light bulb.

Science differs from technology in that science is the knowledge itself while technology is the application of that knowledge to produce material goods. Science and technology are closely related by technological breakthroughs which have given rise to equipment that enables us to perform still more science. In this way, we acquire still more knowledge, which may then be applied to new technologies. As one example, the science that lenses can cause magnification was used for the technology of the microscope. The microscope in turn was used in the science of learning more about tiny organisms.

Science, Technology and Society

Many short examples of technological advances are given. They include such information as the many uses of computer chips, recombinant DNA techniques, agricultural technology and the medical diagnostic tools of magnetic resonance imaging (MRI) and computer tomography (CT).

Chemistry Today: A Closer Look at Some Important Applications

This section focuses on several current chemical advances in more depth. The use of copper-67 for targeting antibodies is described. The application of the chemical method of high temperature combustion for disposing of hazardous wastes such as dioxin is discussed and the **law of six nines** is presented.

Science and Technology: The Benefits and the Risks

This section mentions that hazardous materials or conditions may be formed as byproducts in the course of formulating useful products. It discusses the improper handling of wastes and chemicals which has in turn caused severe problems. It also mentions other incidents which have been the result of poorly handled technology. The chapter concludes with the belief that the safe use of technology should be of concern to everyone. In this country, debates have arisen about where to build nuclear power plants, how to dispose of household wastes and where to allow the smoking of cigarettes, which science has shown to be associated with many health problems. A scientifically informed public can help

3

assess the risks versus the benefits of these and many other technologies for themselves and for their communities.

SELF-TEST

Fill in the Blanks

1. Knowing that water is a liquid at room temperature is a _____ assessment of matter while knowing that it freezes at 32°C is a _____ determination.

2. Obtaining knowledge for the sake of knowledge is called _____, while obtaining knowledge to solve a particular problem is _____.

3. The systematic study of nature is _____ and it involves steps called the _____.

4. An educated explanation about why things happen is called a/an _____ while a/an _____ is a widely accepted explanation.

5. The study of _____ and its _____ is called chemistry.

6. Physics and chemistry are examples of _____.

7. Simple statements or equations about observations of matter are called _____.

8. Using science to produce material goods is called _____.

9. _____was the site of a nuclear accident while _____ and _____ were the sites of hazardous waste dumps and _____ was the site of a severe chemical accident.

10. Trying to find a cure for cancer is an example of _____ science.

11. Dissecting an earthworm to see how its organs are arranged is _____ science.

12. Finding that a laser beam can cut through metal is _____ while using a laser beam for eye surgery is _____.

True or False

1. The use of t-PA to dissolve clots is applied science.

4

2. Theories may be disproved by further experimentation.

3. Phlogiston is needed for combustion.

4. Mass depends upon gravity.

5. Dropping objects to see if they fall tests the law of gravity.

6. Dioxin is used to treat heart patients.

7. The law of six nines states that 99.9999% of a hazard is still present.

8. Checking a light bulb to see if it is burned out is testing a theory.

Short Answer

Tell whether each of the following is science or technology.

1. Baking bread.

2. Finding that copper conducts electricity.

3. Using a match to light charcoal briquettes.

4. Discovering that people suffocate without oxygen.

5. Realizing that caffeine in coffee makes a person nervous.

6. Decaffeinating coffee for drinking.

7. Knowing that water can be found underground.

8. Drilling an oil well for fuel.

9. Listening to a radio.

10. Making a gold bracelet.

11. Seeing that a sunflower follows the sun across the sky.

ANSWERS TO SELF-TEST

Fill in the Blanks

1. qualitative; quantitative
2. basic science; applied science

3. science; scientific method
4. hypothesis; theory   5.   matter; changes
6. natural sciences   7.   laws   8.   technology
9. Chernobyl; Love Canal; Times Beach; Bhopal,India
10. applied   11.   basic
12. basic science, technology

True or False

1. True   2.   True
3. False--combustion requires oxygen
4. False--weight depends on gravity, mass only depends upon amount of material
5. True   6.   False--dioxin is a carcinogen
7. False--it means that 99.9999% of the hazard is gone
8. True

Short Answer

1. technology   2.   science   3.   technology
4. science   5.   science   6.   technology
7. science   8.   technology   9.   technology
10. technology   11.   science

2

ATOMS, MOLECULES, ELEMENTS, COMPOUNDS, AND THE MEASUREMENT OF MATTER

CHAPTER OVERVIEW

This chapter shows the development of chemistry as a science and then introduces many of the terms and problem solving techniques which will be used elsewhere in the course. Chemical and physical properties and changes are discussed. Several different classification schemes for matter are introduced. Dimensional analysis is used as a means of converting between English and metric units and between the metric units themselves. A counting unit called the mole is used to relate atoms on the microscopic scale to grams on the macroscopic scale.

TOPIC SUMMARIES AND EXAMPLES

Introduction

A short history of the roots of chemistry is given. The mystical Egyptian **khemia** is discussed as being applied chemistry. The Greeks were more interested in the basic chemistry of the composition of objects and theorized the existence of the smallest possible particles of matter called **atoms**. The work of the alchemists helped to build a large body of knowledge which was later employed by scientists. Robert Boyle began the use of the scientific method. A scenario gives a view of **deductive reasoning** and states that the making of observations and then drawing conclusions from them is invaluable to the scientific method.

How Modern Chemistry Developed

The efforts of Lavoisier in quantitatively demonstrating the **law of conservation of mass** are presented. This law states that matter is neither created nor destroyed in a chemical reaction, but only changed from one form to another.

Matter is redefined and examples are presented. **Energy** is defined as being the capacity to do work. Heat and mechanical energy are mentioned as being forms of energy.

Matter Explained

Matter can be subdivided by various classification schemes. One way is to divide matter according to its **state** under normal living conditions. The states of matter are **solid, liquid** and **gas**. A second classification is based upon observed properties. Some properties are monitored by **physical changes** which are changes in which the chemical make-up does not change. Other properties are monitored by **chemical changes** which do involve changes in chemical composition. Changes in state, caused by heating or cooling, are physical, rather than chemical, changes because only appearance and not composition changes. In state changes there is only one substance involved instead of the inter-action or the production of two or more substances. **Physical properties** are monitored by the senses or by instruments without chemical changes. **Chemical properties** are determined by chemical changes or attempted chemical changes. A chemical property of iron is that it rusts (reacts with oxygen in the air). A chemical property of gold is that it doesn't rust under the same conditions.

Example 2.1

For each of the following, tell if it is a physical change or a chemical change:

    a) boiling water
    b) cooking an egg in boiling water
    c) cutting a log into two pieces
    d) souring milk
    e) mixing sugar and salt

Solution

    a) physical--cooling produces the original liquid;
        water vapor has the same chemical composition as does
        liquid water
    b) chemical--cooling can't uncook the egg and return it
        to its original appearance; its properties are very
        different from the raw egg
    c) physical--each portion retains the original
        composition and appearance
    d) chemical--sour milk tastes and smells different;
        therefore, new substances have been formed

8

e)   physical--the taste of sugar and that of salt is each
     still apparent in the mixture; neither has formed a
     new substance and changed composition

Example 2.2

    For each of the following, tell if it is a physical
property or a chemical property.

    a)   alcohol boils at 78°C
    b)   charcoal burns in air
    c)   tin conducts electricity
    d)   chlorine is a green gas
    e)   butter becomes rancid if left unrefrigerated

Solution

    a)   physical--cooling returns the original liquid; only
         one substance is involved
    b)   chemical--new substances are produced which look,
         feel and smell different; cooling doesn't restore
         the charcoal to its original appearance
    c)   physical--no composition changes occur when the
         electricity passes through the tin
    d)   physical--color can be observed without even touching
         the object; it undergoes no changes
    e)   chemical--new substances are formed with different
         tastes and smells, indicating a composition change

These properties are related to the subdivision classification scheme of Figure 2-4 in the text. **Heterogeneous** matter
is composed of parts with apparent different properties.
Homogeneous matter has the same appearance and properties
throughout.   Homogeneous matter is subdivided into mixtures
and pure substances.

    Homogeneous mixtures, also called **solutions**, are composed
of parts which can be physically separated.   They have
variable properties which depend upon the relative amounts of
the materials which are mixed. A homogeneous mixture has
similar properties and appearance to one, if not all, of its
constituent parts.   For example, a mixture of salt and water
looks like water but the taste of the salt is still obvious.

    Pure substances cannot be physically subdivided into
anything more simple.   Pure substances include elements and
compounds, each of which have constant compositions and
invariant properties. The 109 known **elements** are the building
blocks of all matter.   They cannot be chemically subdivided
into anything more simple.   The approximately six million
known **compounds** are each composed of two or more elements

9

chemically combined in fixed proportion by mass. All of the compounds can be chemically decomposed to their constituent elements. Compounds are often very different in appearence and properties from the elements from which they are formed. For example, a shiny silver-colored, malleable metal (iron) and a colorless, transparent gas (oxygen) react to form "rust" which is a crumbly red brown solid.

The Atom and the Molecule

An **atom** is the smallest part of an element which retains all the chemical properties of that element. Each element is composed of only one type of atom. A **molecule** is the smallest unit of a compound which retains all the properties of that compound. A molecule is a unique chemical combination of the atoms of two or more elements.

Measuring Matter

We use the metric system of units throughout this course. For the basic units, length is reported in **meters (m)**; mass is reported in **grams (g)**; and volume is reported in **liters (L)**. Larger or smaller units are made by using prefixes which represent powers of ten. Consult Table 2-1 or Appendix B in the main text for a sample of the prefixes and their magnitude. You should have a general idea of the relationship between the English units with which you are familiar and the metric units. See Appendix B for some equivalent values. The **dimensional analysis** method of converting between units is shown in the chapter and is further elaborated upon in Appendix B. This method involves using conversion factors made from equalities to convert from one measuring system to another. Series of these **factor-units** can be employed for complex conversions which involve derived units such as rates and energies or for converting between units for which a single conversion factor cannot be found.

Example 2.3

How many milliliters are in 0.75 liters?

Solution

1000 mL = 1 L so the appropriate factor-units are:

$$\frac{1000 \text{ mL}}{1 \text{ L}} \quad \text{or} \quad \frac{1 \text{ L}}{1000 \text{ mL}}$$

We must use the factor-unit which "cancels out" the unit that we start with (by putting it on the bottom) and leaves the unit that we want (by putting it on the top).

$$? \text{ mL} = (0.75 \text{ L}) \times \frac{1000 \text{ mL}}{1 \text{ L}} = 750 \text{ mL}$$

Example 2.4

How many grams are in 10.0 ounces?

Solution

It is known that 16 oz = 1 lb. Consulting Table B-6 tells us that 1 lb = 454 g. Appropriate factor-units are:

$$\frac{1 \text{ lb}}{16 \text{ oz}} \quad \text{or} \quad \frac{16 \text{ oz}}{1 \text{ lb}} \quad \text{and} \quad \frac{1 \text{ lb}}{454 \text{ g}} \quad \text{or} \quad \frac{454 \text{ g}}{1 \text{ lb}}$$

We can use a series of two factor-units to accomplish the conversion:

$$? \text{ g} = 10.0 \text{ oz} \times \frac{1 \text{ lb}}{16 \text{ oz}} \times \frac{454 \text{ g}}{1 \text{ lb}} = 284 \text{ g}$$

Symbols and Formulas of Elements and Compounds

A **chemical symbol** is the shorthand representation of an element. The **periodic table of the elements** on the front end flaps of the text shows the symbols of the known elements. The back end flaps of the text correlate the symbols with the names of the elements. The symbols are often derived from Greek or Latin names and may not agree with the English name of the element. You should learn the names and symbols of the common elements.

A **chemical formula** is the shorthand representation of a compound. It employs subscripts to show the number and kinds of the atoms of the different elements of which the compound is composed. Foul-smelling formaldehyde, which is used as a preservative for biological specimens, is composed of one atom of carbon, two atoms of hydrogen and one atom of oxygen. Its formula is CH_2O. Sometimes a parentheses is used to set off groups of atoms within a compound. For example, a fertilizer called ammonium phosphate has the formula $(NH_4)_3PO_4$. It has three (NH_4) groups and is composed of three nitrogen atoms, twelve hydrogen atoms, one phosphorus atom and four oxygen atoms.

11

Atomic Weight

The atoms of each element are assigned a **relative atomic weight**. Their masses are compared to the mass of a carbon-12 atom which is assigned a value of 12 atomic mass units (12 amu). Atomic weights are shown below the symbol of the element on the peridic table. We will round these values to whole numbers as is done in the main text.

Example 2.5

Calcium (Ca) has an atomic weight of 40 amu. Bromine (Br) has an atomic weight of 80 amu. Therefore, a Br atom is twice as heavy as a Ca atom. Other atoms can be compared in a similar manner.

Example 2.6

What are the atomic weights of the following elements?

a) U   b) Na   c) Pb   d) Fe

Solution

a) U is 238 amu   b) Na is 23 amu   c) Pb is 207 amu
d) Fe is 56 amu

Gram Atomic Weight and the Mole

Individual atoms are too small to be easily dealt with on a normal basis. It is therefore convenient to talk of a counting unit for large numbers of atoms. A **mole** contains 6.02×10^{23} atoms (Avogadro's number). It is the number of atoms which has a mass in grams numerically equal to the mass of one atom in amu. The mass of one mole (mol) of atoms is called the gram atomic weight.

Example 2.7

1 Ca atom has a mass of 40 amu
6.02×10^{23} Ca atoms has a mass of 40 g
1 mole of Ca atoms has a mass of 40 g

Dimensional analysis techniques can help us to convert between an obtainable mass and this atom counting unit.

Example 2.8

How many moles of lead (Pb) are in 51.8 grams of lead?

12

Solution

The gram atomic weight of lead is 207. This gives us the equality that: 1 mol Pb = 207 g Pb. The factor-units are:

$$\frac{207 \text{ g Pb}}{1 \text{ mol Pb}} \quad \text{and} \quad \frac{1 \text{ mol Pb}}{207 \text{ g Pb}}$$

$$? \text{ moles Pb} = 51.8 \text{ g Pb} \times \frac{1 \text{ mol Pb}}{207 \text{ g Pb}} = 0.250 \text{ mol Pb}$$

Example 2.9

How many grams of silicon (Si) are in 0.75 moles silicon?

Solution

The gram atomic weight of silicon is 28. The factor-units are:

$$\frac{28 \text{ g Si}}{1 \text{ mol Si}} \quad \text{and} \quad \frac{1 \text{ mol Si}}{28 \text{ g Si}}$$

$$? \text{ grams Si} = 0.75 \text{ mol} \times \frac{28 \text{ g Si}}{1 \text{ mol Si}} = 21 \text{ g Si}$$

Formula Weight of a Compound

The formula weight of a compound is the sum of the atomic weights of all the atoms in one molecule of the compound. The gram formula weight of a compound is the mass of one mole of that compound. When a formula weight is found, it can be reported as either amu/formula unit or, more commonly, as gram/mole.

Example 2.10

What is the formula weight of: a) formaldehyde, CH_2O and b) ammonium phosphate, $(NH_4)_3PO_4$?

Solution

a) The atomic weights of C, H and O are 12, 1 and 16, respectively. There are one carbon atom, two hydrogen atoms and one oxygen atom in one molecule of formaldehyde.

The formula weight is:

```
1 x (atomic weight of C) = 1 x 12 = 12
2 x (atomic weight of H) = 2 x  1 =  2
1 x (atomic weight of O) = 1 x 16 = 16
-----------------------------------------
     formula weight of CH_2O   = 30
```

b) The atomic weights of N, H, P and O are 14, 1, 31 and 16, respectively. There are three nitrogen atoms, twelve hydrogen atoms, one phosphorus atom and four oxygen atoms in one formula unit of ammonium phosphate.

The formula weight is:
```
 3 x (atomic weight of N) =  3 x 14 =  42
12 x (atomic weight of H) = 12 x  1 =  12
 1 x (atomic weight of P) =  1 x 31 =  31
 4 x (atomic weight of O) =  4 x 16 =  64
-----------------------------------------
     formula weight of (NH_4)_3PO_4   = 149
```

Compounds and Moles

We can convert between the mass and number of moles of a compound in the same manner as for an element. Note that one mole of a compound is comprised of more than one mole of atoms in the same way that a dozen bicycles has two dozen wheels, one dozen frames, two dozen pedals, etc.

Example 2.11

How many moles of $Ba(OH)_2$ are in 85 g $Ba(OH)_2$?

Solution

First, we must find the gram formula weight. The atomic weights of Ba, O and H are 137, 16 and 1, respectively. The formula weight is 137 + (2 x 16) + (2 x 1) = 171.

$$? \text{ moles } Ba(OH)_2 = 85 \text{ g } Ba(OH)_2 \times \frac{1 \text{ mol } Ba(OH)_2}{171 \text{ g } Ba(OH)_2}$$

$$? \text{ moles } Ba(OH)_2 = 0.50 \text{ mol } Ba(OH)_2$$

Note that the 0.50 mol of $Ba(OH)_2$ would be made of 0.50 mole of barium, 1.0 mole of oxygen and 1.0 mole of hydrogen.

Example 2.12

How many grams are in 0.40 mol $MgBr_2$?

Solution

First, we must find the gram formula weight. The atomic weights of Mg and Br are 24 and 80, respectively. The formula weight is 24 + (2 x 80) = 184.

$$? \text{ grams MgBr}_2 = 0.40 \text{ mol MgBr}_2 \times \frac{184 \text{ g MgBr}_2}{1 \text{ mol MgBr}_2} = 74 \text{ g MgBr}_2$$

SELF-TEST

Fill in the Blanks

1. A solution is a _____ mixture.

2. Pure substances include _____ and _____.

3. _____ is the ability to do work.

4. The ancient Egyptians practiced _____.

5. Matter is neither created nor destroyed in a chemical reaction is a statement of the _____.

6. A/an _____ is a unique chemical combination of a specific number and kind of different atoms.

7. _____ mixtures have nonuniform properties.

8. The metric unit for mass, length and volume are, respectively, _____, _____ and _____.

9. The _____ is the counting number used for atoms and molecules.

10. A/an _____ is the shorthand representation of an element and a/an _____ is the shorthand representation of a compound.

True and False

1. A liter is about the same size as a quart.

2. A sample of gold has a mass of about 0.20 g. That is a chemical property

3. Oil and water form a homogeneous mixture.

4. Magnesium is an element.

5. Deci is the prefix standing for 1/10.

6. The formula $Al_2(CO_3)_3$ represents 6 Al atoms, 3 C atoms and 9 O atoms.

7. A basketball player might have a height of about 2 m.

8. Vinegar is a compound.

9. A freshly poured soft drink is heterogeneous.

10. When leaves decay, it is a chemical change.

11. A Te atom is about twice as heavy as a Cu atom.

12. Compounds always have similar properties to the elements from which they are made.

Short Answer

1. How many amu are in one atom of tin (Sn)?

2. How many grams are in one mole of cobalt (Co)?

3. How many dollars are in one Mdollar?

4. Which is heavier--an atom of gold (Au) or one of silver (Ag)?

5. What is the symbol for:

    a) phosphorus? b) chlorine?   c) sodium?
    d) fluorine?   e) potassium? f) copper?

6. What are the names for:

    a) Fe? b) Ni? c) S? d) Br? e) H? f) Pb?

7. Answer the following with compound, element, heterogeneous mixture or solution.

    a) A transparent yellow liquid seems to have uniform properties. When it evaporates, a yellow residue remains.

    b) A jar is full of white crystals. Not all of the crystals will dissolve in alcohol.

    c) A blue crystal appears to have uniform, consistent properties. It can be broken down electrically to give a silver colored metal and other products.

d) A gas sample is transparent and colorless. The entire mass reacts with sodium in the same way.

e) A brown, transparent liquid has a uniform appearance. As it is boiled, it gets darker and the boiling point changes.

f) A shiny, silvery solid completely turns into a purple gas when heated. When cooled, the gas reforms into the silvery solid. The solid is composed of just one kind of atom.

8. What element has atoms which are about six times as heavy as a sulfur atom?

9. How many of each kind of atom and how many total atoms are in one molecule of:

a) $Cu(NH_3)_4SO_4$ (a bright blue compound)?

b) $C_{15}H_{31}COOC_{30}H_{61}$ (found in beeswax)?

Multiple Choice

1. Which of the following is not a physical property of copper?

a) reddish brown solid
b) melts at $1083^{\circ}C$
c) reacts with nitric acid to make a blue solution
d) conducts electricity

2. Which is a physical change?

a) making vinegar from wine
b) "powdering" sugar
c) burning toast
d) bleaching hair

3. A centimeter is closest in size to:

a) a yard   b) a mile   c) a foot   d) an inch

4. Postage is metered by the ounce. How many grams are in 1.0 ounce?

a) 28 g   b) 7264 g   c) 0.035 g   d) 470 g

5. The length of a football field would be most appropriately measured in:

   a) km   b) mm   c) m   d) cm

6. What is the mass in kilograms of a 180 pound man?

   a) 81720 kg   b) 2520 kg   c) 81.7 kg   d) 396 kg

7. What is the mass of 0.25 mol He?

   a) 0.0625 g   b) 16 g   c) 0.50 g   d) 1.0 g

8. How many moles of Sn are in 5.95 grams of Sn?

   a) 708 mol   b) 0.0500 mol   c) 20.0 mol
   d) 298 mol

9. What is the formula weight of $Mg(CO_2H)_2$?

   a) 58 g/mol   b) 262 g/mol   c) 70 g/mol
   d) 114 g/mol

10. What is the formula weight of sucrose, $C_{12}H_{22}O_{11}$?

    a) 160 g/mol   b) 45 g/mol   c) 81 g/mol
    d) 342 g/mol

11. How many grams are in 4.00 mol $NaHCO_3$ (baking soda)?

    a) 336 g   b) 212 g   c) 0.0476 g   d) 21 g

12. How many moles are in 500 mg $C_9H_8O_4$ (aspirin)?

    a) 360 mol   b) 2.78 mol   c) 0.00278 mol
    d) 90 mol

13. Which contains the most atoms?

    a) 100 g Ca   b) 100 g NH_3 c) 100 g KBr
    d) 100 g $Mg(OH)_2$

14. Which has the largest mass?

    a) 2.0 mol Pb   b) 2.0 mol H_3PO_4
    c) 2.0 mol $Ba(NO_3)_2$ d) all have the same mass

15. Which contains the most atoms?

    a) 1.0 mol Li   b) 1.0 mol S_8 c) 1.0 mol P_4O_{10}
    d) all have the same number of atoms

ANSWERS TO SELF-TEST

Fill in the Blanks

1. homogeneous   2. elements; compounds   3. energy
4. khemia   5. law of conservation of mass
6. molecule   7. heterogeneous
8. gram; meter; liter   9. mole
10. chemical symbol; chemical formula

True or False

1. True--see Table B-6   2. False--physical
3. False--oil floats on top of water   4. True
5. True   6. False--2 Al, 3 C and 9 O
7. True--about 78" or 6'6"
8. False--can have different properties like wine vinegar, cider vinegar, etc.
9. True--gas bubbles and a liquid phase
10. True   11. True--Te is 128 amu and Cu is 64 amu
12. False--think of rust versus oxygen and iron

Short Answer

1. 119 amu   2. 59 g   3. 1,000,000--see Table B-2
4. Au (197 amu) compared to Ag (108 amu)
5. a) P   b) Cl   c) Na   d) F   e) K   f) Cu
6. a) iron   b) nickel   c) sulfur   d) bromine
   e) hydrogen   f) lead
7. a) solution   b) heterogeneous mixture
   c) compound   d) element or compound
   e) solution   f) element
8. Ir--(6 x 32) = 192 amu
9. a) 1 Cu atom, 4 N atoms, 12 H atoms, 1 S atom, 4 O atoms; 22 atoms total
   b) 46 C atoms, 92 H atoms, 2 O atoms; 140 atoms total

Multiple Choice

1. c   2. b   3) d (1 inch = 2.54 cm)
4. a (454/16)   5. c (1 m = 1.09 yard)
6. c (180x454/1000)   7. d (4.0x0.25)
8. b (5.95/119)   9. d (24 + 2x12 + 4x16 + 2x1)
10. d (12x12 + 22x1 + 11x16)   11. a (FW=84; 84x4)
12. c (FW=180; 500x0.001/180)   13. b   14. c   15. c

3

THEORY AND STRUCTURE OF
THE ATOM

CHAPTER OVERVIEW

This chapter deals with the development and history of atomic theory. Experiments which showed the divisibility of the atom are presented. The term radiation is introduced along with its contribution to the evolution of atomic theory. The properties of three subatomic particles and their positions within the atom are explored. The Bohr and quantum mechanical views of the arrangement of electrons within the atom are shown. Also included is the history of development of the periodic chart of the elements. Much of the work that was done was to provide an explanation for the observed repetitiveness of properties of elements. Those properties directly correlate with the number of electrons that the elements possess.

TOPIC SUMMARIES AND EXAMPLES

Introduction

Matter is ultimately composed of atoms. It was originally thought that atoms were indivisible. Studies were then done to see if these tiny atoms were indeed solid or if they were actually composed of still smaller particles.

Dalton's Postulates

Dalton postulated that each element was composed of indivisible, identical, unique atoms that were different from those of all other elements. He theorized that compounds were the result of specific chemical combinations of atoms from two or more different elements. Different arrangements of a set of atoms produced different compounds.

A Historical View of the Atom

Various researchers decomposed compounds to their constituent elements using electricity in a process called **electrolysis**. **Electrolytes** which carry current in solution are attracted to metal rods called **electrodes**. The **anode** with a positive charge attracts ions called **anions** which have a negative charge. The **cathode** with a negative charge attracts **cations** which have a positive charge. In either case, **ions** are atoms which have somehow acquired a charge. This indicated that atoms have some degree of divisiblility and that there is a relationship between matter and electricity.

Cathode Rays and Electrons

In a Crooke's tube, a negatively charged beam which could carry current was found. This negative **cathode ray** beam was independent of the gas or metal used in the tube, indicating that the negative particles were characteristic of all matter. A constant charge-to-mass ratio of the negative **electron** was determined. Oil drop experiments later determined the magnitude of the charge of the electron. By combining the data, the mass of the electron was also found.

The Proton and the Perforated Cathode

In a tube with a perforated cathode, rays were found which travel in the opposite direction from that of the electron. These rays, called **canal rays**, had a positive charge and also had a charge-to-mass ratio which varied with the trace gases present. It appeared that the positive charges "stick together", while the electrons can be lost individually. The highest charge-to-mass ratio was found when hydrogen gas, the gas with the lowest molecular weight, was employed. The smallest possible positive charge was assigned to the **proton**. Although the magnitude of charge of a proton is equal to that of an electron, the mass of the proton (about 1 amu) is 1837 times larger than that of an electron.

J.J. Thomson's View of the Atom

This "plum pudding" model of electrons embedded in clouds of protons has been discredited.

Finding Order Among Atoms--The Development of the Periodic Chart

Patterns are found in both chemical and physical behavior. As atomic weights increase, certain properties, such as state and the type of compounds formed, recur. The first periodic charts were organized to reflect these trends.

21

The horizontal rows, called **periods**, had increasing atomic weight. The vertical columns, called **groups** or **families**, have similar properties. It was thought that the atomic structure was related to the number of subatomic particles within the atom. A method of correlating those particles with the observed properties was desired and further experiments were performed to try to discover the relationship.

Radioactivity

It was found that penetrating rays called **X-rays** were given off by cathode ray tubes. Other substances gave off **radiation** without being triggered by electricity. **Radioactivity** is the ability to produce penetrating radiation without the input of electricity or other energy. The most frequently observed radiations are classified as **alpha particles, beta particles** and **gamma rays**. Of these three emissions, alpha particles are the most massive while gamma rays are the most penetrating. The emissions could be used as particle guns to explore the interior of the atom.

Rutherford and the Nuclear Atom: A First for the Twentieth Century

Alpha particles, which have a mass of about 4 amu, were directed at gold foil and a small percentage were deflected. This gave rise to the **nuclear** model of the atom. The **nucleus** is the center of charge and is positive. It contains the protons and, thus, most of an atom's mass, in a volume which is about 1/100,000 that of the atom itself.

Example 3.1

To get an idea of the relationship of the size of the nucleus to that of the atom, visualize a human hair placed on the fifty-yard line of a football field. The width of the hair has about the same relationship to the length of the playing field as the diameter of a nucleus does to the diameter of an atom.

The bulk of the volume of the atom is occupied by the electrons which are outside the nucleus and offer little resistance to a massive alpha particle.

The Neutron: Another Subatomic Particle

Discovery of the **neutron**, which has no charge and a mass slightly larger than that of a proton, completed the picture of the atom and accounts for the "missing mass". Neutrons are also found in the nucleus. The properties of the three subatomic particles are summarized in Table 3-2 of the text.

Atomic Number--Numbering the Elements

The **atomic number** is the number of protons in an atom and appears above the symbol of the element on the periodic chart. The atomic number defines the atom. In other words, if an atom has 16 protons, it must be sulfur and, if it is sulfur, it must have 16 protons. In a neutral atom, the number of electrons must be equal to the number of protons.

The Existence of Isotopes

The atoms of most of the elements are not all identical because of the existence of isotopes. **Isotopes** of an element have different numbers of neutrons and, thus, different masses. The number of neutrons plus the number of protons is called the **mass number**. Specific isotopes are designated by the name of the element plus the mass number. For example, carbon-12 has a mass number of 12 while carbon-13 has a mass number of 13. Each has 6 protons since the atomic number of carbon is 6. Carbon-12 has 12-6 = 6 neutrons and carbon-13 has 13-6 = 7 neutrons. The **isotopic notation** of an element shows both the mass number as a superscript and the atomic number as a subscript. The two carbon isotopes mentioned earlier would be:

$$^{12}_{6}C \text{ (carbon-12)} \quad \text{and} \quad ^{13}_{6}C \text{ (carbon-13)}$$

Example 3.2

List the name of the isotope, the atomic number, the mass number, the number of protons (p^+), electrons (e^-) and neutrons (n) for the following isotopes.

a) $^{11}_{5}B$ b) $^{38}_{18}Ar$ c) $^{109}_{47}Ag$ d) $^{149}_{62}Sm$ e) $^{208}_{82}Pb$

Solution

Remember that in a neutral atom the number of electrons must equal the number of protons and the the mass number is the sum of the neutrons and the protons.

	Name of isotope	Atomic #	Mass #	#p^+	#e^-	#n
a)	boron-11	5	11	5	5	6
b)	argon-38	18	38	18	18	20
c)	silver-109	47	109	47	47	62
d)	samarium-149	62	149	62	62	87
e)	lead-208	82	208	82	82	126

The **atomic weight** of an element is the average of the

masses of all the isotopes of that element weighted by the isotopic abundance of each isotope on earth. See Table C-1 in the appendix for a list of some of the naturally occurring isotopes and their natural abundances.

Example 3.3

Bromine exists in nature as two isotopes and is about 50.5% bromine-79 and 49.5% bromine-81. To find the approximate atomic weight, the decimal percent can be multiplied by the mass number and the results summed.

$$
\begin{array}{r}
0.505 \times 79 = 39.9 \\
+ \quad 0.495 \times 81 = 40.1 \\
\hline
\end{array}
$$

80.0 = approximate atomic weight

This is only an approximation of the atomic weight which is shown on your text because the actual mass of each isotope is not the same as the mass number.

The Bohr Model of the Atom

Different colors of light have different energies. Each element produces a characteristic "fingerprint" **line spectrum** indicating that it has certain specific energies associated with it. Bohr proposed that electrons were positioned in certain orbits or **shells** which represent specific energy levels (represented by **n**) at different distances from the nucleus. Electrons can move to higher energy levels when they are **excited** by heat or electricity. When they return to the lower energy level or **ground state**, they release energy in a **quantum** of light energy which is related to the difference in energy between the two levels. Each transition between energy levels gives rise to a specific color. (**Ionization** occurs when an electron absorbs sufficient energy to totally escape the atom.) The observed line spectrum of hydrogen could be explained by this theory, which was then extrapolated to describe the arrangement of electrons in other atoms.

Example 3.4

Although this discussion of the atom may not seem pertinent to daily life, you have seen the effects of electrons moving between different energy levels. The continuous spectrum of visible light appears in a rainbow or in sunlight shining through a prism. Neon lights, fireworks and "colored" fireplace logs are all examples of the specific colors given off when excited electrons return to lower energy levels within specific atoms.

Applying Bohr's Atomic Model

Bohr gave letters and numbers to each of the energy levels. In order of increasing energy, they are K (n = 1), L (n = 2), M (n = 3), N (n = 4), etc. Each shell can hold $2n^2$ electrons, so K can contain $2(1)^2$ = 2 electrons, L can contain $2(2)^2$ = 8 electrons, etc.

Electronic Configurations and the Bohr Model

An **electronic configuration** is a description of the occupied energy levels in an atom. Each level fills until all electrons are used up. This system works well for the first eighteen elements. The **octet rule** or the extreme stability of the **noble gases** is mentioned as being related to the discontinuity in which the next level begins to fill before the previous one is full. See Table 3-3 in the text. The outermost electrons, which are those beyond the ones contained in the previous noble gas, are found to control the chemical and physical properties observed for the elements.

Example 3.5

What are the Bohr electronic configurations of:

a) carbon (C); and b) phosphorus (P)?

Solution

a) carbon has atomic number 6; it has 6 electrons
   K2, L4 is the Bohr configuration
b) phosphorus has atomic number 15; it has 15 electrons
   K2, L8, M5 is the Bohr configuration

The Quantum Mechanical Model of the Atom

The Bohr theory is not able to adequately explain the electron energies and locations of atoms of higher atomic number than Ar. The **quantum mechanical model of the atom** utilizes the idea that electrons have wave behavior as well as particle behavior. Instead of having defined orbits, electrons have **probabilities** of being in certain locations called **orbitals**. An orbital represents a region of space in which an electron has a 95% probability of being found. There are four types of orbitals, also called sublevels, found in the ground states of known atoms. The maximum capacity of the levels are: s--2 electrons, p--6 electrons, d--10 electrons and f--14 electrons. The first energy level has only the s sublevel. The second energy level has both s and p sublevels. The third energy level has s, p and d sublevels. The fourth energy level contains all four sublevels.

Inspection of the periodic chart of the elements reveals that they fall neatly into: 1) a left hand block which is two elements wide; 2) a right hand block which is six elements wide; 3) a central section which is ten elements wide; and 4) a section at the bottom (which is removed from numerical sequence for the purpose of compactness) that is fourteen elements wide. These blocks on the periodic chart correspond to the **s, p, d** and **f** sublevel designations. More detailed electronic configurations than those of Bohr can be written using a quantum mechanical notation. In this notation, the energy level is used as a coefficient followed by the sublevel designation. The number of electrons in that sublevel is then shown as a superscript. The periodic chart or Figure 3-32 in the text can be used to predict the filling order. The sublevels may not be overfilled and the sum of the superscripts must equal the number of electrons in the atom. The outermost electrons which are responsible for an element's properties are again designated as those being beyond the configuration associated with the previous noble gas.

Example 3.6

What are the electronic configurations for:

a) Al   b)   Ni   c)   Sn

Solution

a)   13 electrons--$1s^2 2s^2 2p^6 3s^2 3p^1$

b)   28 electrons--$1s^2 2s^2 2p^6 3s^2 3p^6 4s^2 3d^8$

c)   50 electrons--$1s^2 2s^2 2p^6 3s^2 3p^6 4s^2 3d^{10} 4p^6 5s^2 4d^{10} 5p^2$

SELF-TEST

Fill in the Blanks

1. Elements are arranged on the periodic chart in horizontal rows called _____ which have _____ and in vertical _____ which have _____.

2. Gaseous elements give _____ when heated.

3. Charged particles are called _____.

4. A cathode has a/an _____ charge.

5. _____ give the probability of finding an electron in a certain volume of space.

26

6. _____ determined the charge on an electron.

7. The **d** sublevel can hold _____ electrons.

8. _____ devised the nuclear model of the atom.

9. Natural radiations include _____, _____ and _____.

10. Isotopes of an element have the same number of _____ and _____, but different numbers of _____.

11. Elements in the same family of the periodic chart have the same number of _____ electrons.

12. _____ rays have a positive charge while _____ rays are negative.

True or False

1. Electrolytes carry current in solution.

2. F and Cl have the same number of outermost electrons.

3. The O energy level can hold 50 electrons.

4. A Ca atom has two **d** electrons.

5. The electronic configuration of F is:
   $1s^2 2s^2 2p^6 3s^2 3p^6 4s^1$.

6. A neutron is slightly more massive than a proton.

7. Canal rays always have the same charge-to-mass ratio.

8. An alpha particle is a more penetrating radiation than a gamma ray.

9. The ground state of a Zn atom has ten **d** electrons.

10. Atoms are divisible.

11. Mass number is the sum of the neutrons, electrons and protons in an atom.

12. An Os atom should have 190 electrons in its electronic configuration.

Short Answer

1. Answer the following with the correct subatomic particles--protons, electrons and/or neutrons.

   a) found in the nucleus
   b) responsible for chemical behavior
   c) positively charged
   d) lightest in mass
   e) last discovered
   f) equals the atomic number at all times
   g) equal in a neutral atom
   h) determines the name of the element
   i) causes differences in the mass of the atoms of a particular element

2. Fill in a table with the name, the mass number, the atomic number and the number of protons, electrons and neutrons in the following neutral atoms.

   a) $^{22}_{10}Ne$ b) $^{43}_{20}Ca$ c) $^{53}_{24}Cr$ d) $^{80}_{34}Se$

   e) $^{116}_{48}Cd$ f) $^{171}_{70}Yb$ g) $^{196}_{80}Hg$ h) $^{205}_{81}Tl$

3. Write the name and symbol for the following isotopes.

   a) 3 p^+, 4 n   b) 16 p^+, 20 n   c) 28 p^+, 36 n
   d) 36 p^+, 57 n   e) 56 p^+, 82 n   f) 74 p^+, 116 n

4. Give the Bohr configuration (K, L, M, etc.) for the following elements.

   a) B   b) Mg   c) S   d) Li   e) Al

5. Write the quantum mechanical electronic configuration ($1s^2$, etc.) for the following elements.

   a) Be   b) F   c) Na   d) Ca   e) V   f) Ge
   g) Br   h) Tc   i) Xe   j) Pt   k) Ra

Multiple Choice

1. Which pair below has the same number of outermost electrons?

   a) N, O   b) B, Al   c) K, Sr   d) Sc, B

2. How many L electrons are in a Si atom?

    a) 2   b) 4   c) 6   d) 8

3. Which species has the largest number of neutrons?

    a) $^{164}_{66}Dy$ b) $^{165}_{67}Ho$ c) $^{166}_{68}Er$

    d) all have the same number of neutrons

4. Not all elements have an electronic configuration which follows the pattern shown in the text. What element has the electronic configuration below?

    $1s^2 2s^2 2p^6 3s^2 3p^6 4s^1 3d^5$

    a) Mn   b) Mg   c) Cr   d) none of these fit

5. How many outermost electrons are in a Cl atom?

    a) 0   b) 2   c) 5   d) 7

6. Silicon is expected to have properties most like those of which element below?

    a) C   b) Al   c) P   d) N

7. Which choice includes a pair of elements in the same period?

    a) Se, Te   b) Mg, B   c) P, Cl   d) Ar, Kr

8. Which species has the largest number of protons?

    a) iron-54   b) cobalt-59   c) manganese-55
    d) nickel-58

9. Which species has the largest number of neutrons?

    a) iron-54   b) cobalt-59   c) manganese-55
    d) nickel-58

ANSWERS TO SELF-TEST

Fill in the Blanks

1. periods; increasing atomic number; groups or families; similar properties
2. line spectra   3. ions   4. negative
5. orbitals   6. Millikan   7. 10   8. Rutherford
9. alpha; beta; gamma

10. protons; electrons; neutrons
11. outermost   12. canal; cathode

True or False

1. True
2. True--each has 7 electrons past the noble gas configuration
3. True--$2(5)^2 = 50$ 4. False--$1s^2 2s^2 2p^6 3s^2 3p^6 4s^2$
5. False--F has 9 not 19 electrons 6. True
7. False--depends on the gases involved
8. False--the reverse is true
9. True--$1s^2 2s^2 2p^6 3s^2 3p^6 4s^2 3d^{10}$
10. True--can be broken into p^+, e^- and n
11. False--only neutrons plus protons
12. False--only 76 electrons; 190 is the atomic weight

Short Answer

1. a) neutrons and protons   b) electrons
    c) protons   d) electrons   e) neutrons
    f) protons   g) protons and electrons   h) protons
    i) neutrons

	Name of isotope	Atomic #	Mass #	#p^+	#e^-	#n
2. a)	neon-22	10	22	10	10	12
b)	calcium-43	20	43	20	20	23
c)	chromium-53	24	53	24	24	29
d)	selenium-80	34	80	34	34	46
e)	cadmium-116	48	116	48	48	68
f)	ytterbium-171	70	171	70	70	101
g)	mercury-196	80	196	80	80	116
h)	thallium-205	81	205	81	81	124

3. a) lithium-7, ^7_3Li b) sulfur-36, $^{36}_{16}\text{S}$

    c) nickel-64, $^{64}_{28}\text{Ni}$ d) krypton-93, $^{93}_{36}\text{Kr}$

    e) barium-138, $^{138}_{56}\text{Ba}$ f) tungsten-190, $^{190}_{74}\text{W}$

4. a)   5 electrons, K2, L3
    b) 12 electrons, K2, L8, M2
    c) 16 electrons, K2, L8, M6
    d)   3 electrons, K2, L1
    e) 13 electrons, K2, L8, M3

5. a) 4 electrons, $1s^2 2s^2$

   b) 9 electrons, $1s^2 2s^2 2p^5$

   c) 11 electrons, $1s^2 2s^2 2p^6 3s^1$

   d) 20 electrons, $1s^2 2s^2 2p^6 3s^2 3p^6 4s^2$

   e) 23 electrons, $1s^2 2s^2 2p^6 3s^2 3p^6 4s^2 3d^3$

   f) 32 electrons, $1s^2 2s^2 2p^6 3s^2 3p^6 4s^2 3d^{10} 4p^2$

   g) 35 electrons, $1s^2 2s^2 2p^6 3s^2 3p^6 4s^2 3d^{10} 4p^5$

   h) 43 electrons, $1s^2 2s^2 2p^6 3s^2 3p^6 4s^2 3d^{10} 4p^6 5s^2 4d^5$

   i) 54 electrons, $1s^2 2s^2 2p^6 3s^2 3p^6 4s^2 3d^{10} 4p^6 5s^2 4d^{10}$
       $5p^6$

   j) 78 electrons, $1s^2 2s^2 2p^6 3s^2 3p^6 4s^2 3d^{10} 4p^6 5s^2 4d^{10}$
       $5p^6 6s^2 4f^{14} 5d^8$

   k) 88 electrons, $1s^2 2s^2 2p^6 3s^2 3p^6 4s^2 3d^{10} 4p^6 5s^2 4d^{10}$
       $5p^6 6s^2 4f^{14} 5d^{10} 6p^6 7s^2$

Multiple Choice

1. b (B-L3 and Al-M3) 2. d (L shell is full)

3. d 4. c (Cr has 24e$^-$) 5. d ($3s^2 3p^5$ or M7)
6. a (same group) 7. c (same horizontal row)
8. d (Ni has 28 protons) 9. b (Co-59 has 32 neutrons)

4

CHEMICAL BONDING: HOLDING MATTER TOGETHER

CHAPTER OVERVIEW

This chapter discusses the ionic and covalent bonding of elements to form compounds. Electronegativity is used to determine the type of bonding and the polarity of covalent bonds. The three dimensional structure of a molecule is needed to determine whether it is polar or nonpolar. The concept of oxidation numbers is introduced. Some common monatomic and polyatomic ions are tabulated. Formulas are written from the names of compounds and the names of compounds are determined from their formulas.

TOPIC SUMMARIES AND EXAMPLES

Introduction

Most elements are found in nature as parts of compounds rather than as the uncombined element. This is because most elements have electronic configurations that cause them to be more stable in the combined form than alone.

Lewis Dot Diagrams

Elements in the same group have similar properties because they have the same number of outermost or **valence** electrons beyond those of the previous noble gas. These valence electrons can be given up, accepted or shared and they determine the element's chemical and physical properties. A **Lewis Dot Diagram** of an element shows the symbol of the element as representing the nucleus and all electrons in closed shells. **Dots** stand for the valence electrons. These can be determined by writing out the electronic configuration or by noting that the group number provides the number of valence electrons for the "A" group elements. (See Figure 4-1

32

in the text.)

Example 4.1

Write the Lewis Dot Diagram for the following elements.

a) Ca   b) S   c) N   d) H

Solution

a) Ċa·   b) ·S̈:   c) ·N̈·   d) H·

Covalent Bonding

The noble gas structure has been found to be very stable. Other elements have a tendency to try to attain that same stability. One way to do that is to gain a share in eight electrons to obtain the stable **octet** seen in all of the noble gases except helium. Hydrogen gains stability by sharing two electrons to be similar to the closed first energy level of helium. A **covalent bond** is comprised of the shared electrons. Electrons associated with the atom which are not shared with another atom are called **unshared** electrons.

The Diatomic Elements

Figure 4-2 in the text presents the Lewis dot diagrams for the **diatomic elements** (H_2, N_2, O_2, F_2, Cl_2, Br_2 and I_2). Note that the hydrogen atoms share two electrons. In each of the other cases, the sum of the shared electrons plus the unshared electrons equals eight for each atom. One pair of electrons shared constitutes a **single covalent bond** also simply called a single bond. Two pairs shared is a **double bond**. Three shared pairs is a **triple bond**. No ordinary compound contains bonds in which more than three pairs of electrons are shared between two atoms. Each complete dot structure should account for no more and no fewer than all of the available valence electrons. Sometimes dashes are used to show shared electron pairs. On other occasions, the unshared electrons are omitted from the picture. This is particularly useful in the case of more complex molecules.

Other Molecules Having Covalent Bonds

When two or more different elements bond, the octet rule still must be followed. The structures should be made to appear as symmetrical as possible. In multiple atom species, the "odd" atom is usually the central one. In some cases, more than one Lewis dot structure which appears to follow the octet rule may be drawn. This occurs for compounds in which the real arrangement of the electrons is not adequately described by the Lewis diagram technique.

Example 4.2

Write the Lewis dot diagrams and dash notations for the compounds below.

    a) AsH_3 b) CS_2 c) HBr d) SiF_4

Solution

    a) As has 5 valence electrons and each H has 1 for a total of 8 valence electrons

```
      ..                    ..
 H :As: H            H —As —H
      ..                     |
      H                      H
```

    b) C has 4 valence electrons and each S has 6 for a total of 16 valence electrons

```
 ..       ..              ..         ..
 S :: C :: S             S  = C =  S
 ..       ..              ..         ..
```

For reasons of symmetry, this is preferred to:

```
 ..
 :S : C:::S:
 ..
```

which also follows the octet rule.

    c) H has 1 valence electron and Br has 7 for a total of 8 valence electrons

```
      ..                    ..
 H : Br:             H — Br:
      ..                    ..
```

    d) Si has 4 valence electrons and each F has 7 for a total of 32 valence electrons

```
        ..                      ..
       :F:                     :F:
  ..   ..  ..             ..    |   ..
 :F : Si : F:            :F — Si — F:
  ..   ..  ..             ..    |   ..
       :F:                     :F:
        ..                      ..
```

Ionic Bonding

In ionic bonding, electrons are transferred to create octets. The ions which are formed have configurations which are the same as a noble gas. The atom which loses electrons becomes positive creating a **cation**. The atom which gains electrons becomes negative creating an **anion**. These ions are held together by their opposite charges. The resulting

attraction is called an **ionic bond**. Remember that only
electrons can be transferred. The nucleus with its protons
and neutrons is <u>not</u> involved in chemical reactions.

Example 4.3

   What are the ionic compounds formed from: a) Sr and S
and b) K and O?

Solution

   a) Sr has a dot diagram of Sr·

       If it loses 2 e⁻, it will have 12-2 = 10 e⁻ and the
       same configuration as the stable Kr (8 outermost e⁻).

       S has the dot diagram ·S̈:

       If it gains 2 e⁻, it will have 16+2 = 18 e⁻ and the
       same configuration as the stable Ar (8 outermost e⁻).

       Sr· + ·S̈: --> Sr^{2+} :S̈: $^{2-}$ (the compound is SrS)

   b) K has the dot diagram K·

       If it loses 1 e⁻, it will have 19-1 = 18 e⁻ and the
       same configuration as the stable Ar (8 outermost e⁻).

       O has the dot diagram ·Ö:

       If it gains 2 e⁻, it will have 8+2 = 10 e⁻ and the
       same configuration as the stable Ne (8 outermost e⁻).

       K· + ·Ö: --> K^+ :Ö: $^{2-}$ (the compound is K_2O)
       K· K^+

**Using Electronegativity to Determine Whether a Bond is Ionic
or Covalent**

   Electronegativity (EN) is a measure of an element's
relative attraction for electrons in a bond. Table 4-1 in the
text has a list of electronegativities. Table 4-2 in the text
correlates the difference in electronegativity between two
atoms to the relative percent ionic/percent covalent character
of the bond. In general, a bond is considered to be ionic if
the electronegativity difference between the two atoms is
greater than 1.6. If you study the examples here and in the
text, you will note that in general: 1) bonds between two
nonmetals tend to be covalent; and, 2) bonds between a metal
and a nonmetal tend to be ionic.

Example 4.4

Tell if the following bonds are ionic or covalent.

a) HCl   b) CsBr   c) CO   d) AlF$_3$

Solution

a) H (EN = 2.1), Cl (EN = 3.0);
    EN difference is 0.9, therefore considered covalent
b) Cs (EN = 0.7), Br (EN = 2.8);
    EN difference is 2.1, therefore considered ionic
c) C (EN = 2.5), O (EN = 3.5);
    EN difference is 1.0, therefore considered covalent
d) Al (EN = 1.5), F (EN = 4.0);
    EN difference is 2.5, therefore considered ionic

The Polar and Nonpolar Covalent Bond

When the atoms sharing electrons have the same electronegativity, the bond is said to be a **nonpolar** covalent bond. When the electronegativities of the bonded atoms are different, the bond is said to be a **polar** covalent bond (up to the electronegativity difference at which the bond is considered to be ionic). Greater electronegativity differences mean more polar bonds.

Example 4.5

Rank the following bonds in order of increasing polarity.

H$_2$S   O$_3$ NH$_3$ ICl$_3$

Solution

H$_2$S: H (EN = 2.1), S (EN = 2.5); EN difference = 0.4
O$_3$:   O (EN = 3.5); EN difference = 0
NH$_3$:   N (EN = 3.0), H (EN = 2.1); EN difference = 0.9
ICl$_3$: I (EN = 2.5), Cl (EN = 3.0); EN difference = 0.5

O$_3$ H$_2$S   ICl$_3$ NH$_3$
----------------------> increasing polarity

The Three-Dimensional Characteristics of Molecules: Some are Polar and Some are Not

If a molecule has no polar bonds, it is a **nonpolar** molecule. When a molecule has polar bonds, it may or may not be polar. If polar bonds are asymmetrically aligned so that the center of positive charge does not coincide with the center of negative charge, the molecule is a **polar molecule**. If the polar bonds are symmetrically arranged so that the center of positive charge does coincide with the center of negative charge, the molecule is a **nonpolar molecule**.

Example 4.6

Are the following molecules polar or nonpolar?

a) CO (linear)   b) H_2S (bent)
c) $SiBr_4$ (tetrahedral)   d) N_2 (linear)

Solution

a) This has one polar bond. Since it is asymmetrical, the molecule is polar.
b) This has two polar bonds. The bent structure causes the molecule to be asymmetrical and it is polar.
c) This molecule has four polar bonds. They are symmetrically arranged, so the molecule is nonpolar.
d) There are no polar bonds. The molecule is nonpolar.

How to Write a Chemical Formula

Oxidation numbers are the real charges of ions in ionic compounds and the charges assigned to atoms in covalent compounds. Any free element has an oxidation number of zero. The algebraic sum of the oxidation numbers in a compound must be zero. In other words, the sum of the positive charges must be equal in magnitude to the sum of the negative charges. Metals always have positive oxidation numbers. When two nonmetals are bonded, the more electronegative element always has its normal negative charge and the other is assigned the appropriate positive charge.

Example 4.7

What is the oxidation number of the underlined element in each compound?

a) $\underline{Ti}S_2$ b) $\underline{Mn}_2O_3$ c) $\underline{Cu}_3P$ d) $\underline{U}F_6$ e) $\underline{S}O_3$

Solution

a) S is -2 so Ti must be +4 [1(4) + 2(-2) = 0]
b) O is -2 so Mn must be +3 [2(3) + 3(-2) = 0]
c) P is -3 so Cu must be +1 [3(1) + 1(-3) = 0]
d) F is -1 so U must be   +6 [1(6) + 6(-1) = 0]
e) O is more electronegative than S. Therefore, it is assigned the -2 charge so S is +6. [1(6) + 3(-2) = 0]

Table 4-3 in the text lists the names, formulas and charges of important monatomic and polyatomic ions. Many of the charges of the monatomic ions can be determined from the periodic chart, however, the polyatomic ions must be memorized. For metals which exhibit only one oxidation number (primarily the "A" group metals), the metal name is used with the name of the appropriate anion. Variable valence metals, which can have more than one oxidation state, have in their names a Roman numeral to indicate the oxidation state in that particular compound. The formulas to go with the names can be written easily if you remember that the sum of the oxidation numbers must be zero.

Example 4.8

Write the formulas for the following compounds.

a) calcim bromide   b) potassium phosphide
c) copper(II) nitrate   d) gold(III) phosphate
e) aluminum fluoride   f) tin(IV) sulfite
g) barium acetate   h) magnesium arsenate

Solution

a) Ca^{2+} and Br^- becomes $CaBr_2$

b) K^+ and P^{3-} becomes K_3P

c) Cu^{2+} and NO_3^- becomes $Cu(NO_3)_2$

d) Au^{3+} and PO_4^{3-} becomes $AuPO_4$

e) Al^{3+} and F^- becomes AlF_3

f) Sn^{4+} and SO_3^{2-} becomes $Sn(SO_3)_2$

g) Ba^{2+} and $C_2H_3O_2^-$ becomes $Ba(C_2H_3O_2)_2$

h) Mg^{2+} and AsO_4^{3-} becomes $Mg_3(AsO_4)_2$

How to Write a Chemical Name

For a constant valence metal, the metal name is written first, followed by the name of the ion to which it is bonded. For a variable valence metal, the metal name is written first, followed by its oxidation number in that compound and, finally, the name of the ion to which the metal is bonded. When two nonmetals are bonded, the less electronegative element is named first, followed by the ion name of the more electronegative element. Greek prefixes, listed in the book, are used to tell how many atoms of each nonmetal are in the formula of the molecule.

Example 4.9

What are the names of the following compounds?

a) $(NH_4)_3P$ b) $Sr_3(PO_4)_2$ c)   FeS   d) $AsCl_3$
e) CS_2 f) $BeCO_3$ g) Li_2CrO_4 h) $V(NO_2)_4$

Solution

a) ammonium phosphide   b)   strontium phosphate
c) iron(II) sulfide   d)   arsenic trichloride
e) carbon disulfide   f)   beryllium carbonate
g) lithium chromate   h)   vanadium(IV) nitrite

SELF-TEST

Fill in the Blanks

1. Prefixes are used in naming compounds composed of
   _____.

2. If the electronegativity difference between two atoms
   is greater than 1.6, the bond is considered to be
   _____.

3. Cations are formed when electrons are _____.

4. In a double covalent bond, _____ electrons are
   shared.

5. Nitrogen has _____ valence electrons.

6. Polar covalent bonds are formed when the two atoms
   bonded have different _____.

7. Roman numerals are used in naming compounds when the
   metal can have more than one _____.

8. The elements which are very stable because they have
   eight electrons in their outer shell are _____.

9. Bonds between the atoms in any of the diatomic
   elements are _____ covalent bonds.

10. When nonmetals are in ionic compounds, their ions
    always have a _____ charge.

11. A sulfide ion has a charge of _____.

True or False

1. A Na^+ ion has 12 protons and 11 electrons.

2. Sulfur and oxygen combine to make covalent compounds.

3. Triple bonds have three electrons shared.

4. Hydrogen shares only two electrons in covalent bonds.

5. All noble gases have a stable octet.

6. The stable ion formed from Ca has 18 total electrons.

7. Ba and F react to form an ionic compound.

8. All of the VIIA elements are expected to form -1 ions in ionic compounds.

9. The ion formed when two electrons are given up has a -2 charge.

10. Asymmetrical polar bonds cause a molecule to be polar.

Short Answer

1. Give the Lewis dot diagrams for:

   a) Sr   b) Rb   c) Al   d) Se   e) As

2. Draw the Lewis dot diagrams for:

   a) H_2SO_4 (O's bonded to S, H's to O's)
   b) $SeBr_2$
   c) $HClO_2$ (H bonded to O, O's bonded to Cl)
   d) H_2CS (H's and S bonded to C   e) SO_3
   f) ICl   g) $AsBr_3$ h) HSCN (linear)
   i) CBr_4 j) H_2Se
   k) CH_3OH (3 H's and the O bonded to the C, the other H bonded to the O
   l) C_2Cl_2 (C's bonded together, Cl's on different C's)

3. Tell if the following compounds are polar or nonpolar.

   a) SO_2 (bent)   b) O_3 (bent)   c) HBr (linear)
   d) CS_2 (linear)   e) SiH_4 tetrahedral
   f) PCl_3 (tetrahedral)
   g) CH_3Cl (tetrahedral, H's and Cl bonded to C)
   h) H_2

40

4. What are the formulas and the names of the ionic compounds formed from the following elements?

a) Ca and N   b) Fr and P   c) Na and I
d) Li and O   e) Mg and Cl   f) Rb and N
g) Al and O   h) Ba and Br

5. Tell if the following bonds between atoms are classified as primarily covalent or primarily ionic.

a) C and S   b) P and I   c) Na and O
d) S and O   e) Cs and S   f) As and H
g) Li and Cl   h) F and I

6. What is the oxidation number of the underlined element in each compound below?

a) $\underline{P}Cl_3$ b) $\underline{Fe}_2O_3$ c) $\underline{Cu}S$ d) $\underline{Cl}_2O_7$
e) $\underline{Mn}Br_4$ f) $\underline{Sc}P$ g) $\underline{Sn}O_2$ h) $\underline{N}_2O_5$
i) $\underline{Ti}_3N_4$ j) $\underline{Pb}S$ k) $\underline{S}F_6$ l) $\underline{Ni}_3P_2$

7. Give the correct formulas for the followinng names.

a) iron(III) fluoride   b) mercury(II) oxide
c) iodine pentafluoride   d) sodium phosphide
e) lithium chloride   f) magnesium nitrite
g) barium phosphate   h) tin(IV) acetate
i) dinitrogen pentoxide   j) phosphorus trichloride
k) cobalt(III) chromate   l) strontium carbonate
m) nickel(III) hydroxide   n) vanadium(V) sulfide
o) uranium(VI) phosphate   p) bismuth(III) sulfate
q) potassium iodide   r) carbon tetrabromide

8. Give the correct names for the following formulas.

a) $TiCl_4$ b) SF_6 c) K_3N d) WO_3
e) CuF f) SiO_2 g) IF_3 h) N_2O_4
i) Na_2S j) $Mn(SO_3)_2$ k) $Fe(NO_3)_2$ l) NO
m) Be_3N_2 n) $Sr(OH)_2$ o) $U(NO_2)_3$ p) $CoAsO_4$
q) PbO_2 r) B_2O_3

Multiple Choice

1. Which pair of elements has the same number of valence electrons?

a) Mg, Al   b) Mn, F   c) O, Se   d) K, Ar

2. Which compound below should be the most ionic?

a) SrO b) CO c) P_4O_{10} d) SiO_2

3. Which compound below should be the most covalent?

    a) K_2S b) MgS   c) Al_2S_3 d) CS_2

4. Which is necessary to have polar bonds?

    a) high atomic number   b) 8 electrons
    c) fluorine   d) different electronegativies

5. Which is the incorrect formula for a compound?

    a) $RbNO_3$ b) Al_2S_3 c) $Ca_2(PO_4)_3$ d) $(NH_4)_2S$

6. Which is the correct name for the given compound?

    a) $CaBr_2$, calcium(II) bromide
    b) P_4O_5, pentaphosphorus tetroxide
    c) $FeCl_3$, iron(III) chloride
    d) MnO_2, manganese dioxide

7. How many total electrons must be shared to satisfy the octet rule in PI_3?

    a) 2   b) 4   c) 6   d) 8

8. How many total electrons must be shared to satisfy the octet rule in SeO_3?

    a) 2   b) 4   c) 6   d) 8

9. Which of the following compounds has a double bond?

    a) C_2H_4 b) NCl_3 c) OF_2 d) ICl

10. Which polyatomic ion has a 3- charge?

    a) chromate   b) phosphate   c) nitrite
    d) sulfate

ANSWERS TO SELF-TEST

Fill in the Blanks

    1. two nonmetals 2. ionic 3. given up 4. four
    5. five 6. electronegativities 7. oxidation state
    8. noble gases 9. nonpolar 10. negative 11. 2-

True or False

    1. False--electrons, not protons, are given up or
        accepted

2. True   3. False--three pairs or 6 e$^-$
4. True   5. False--He is stable with 2 e$^-$
6. True--loses 2 e$^-$ to be like Ar   7. True   8. True
9. False--is left with 2 extra protons, so has +2 charge
10. True

Short Answer

1. 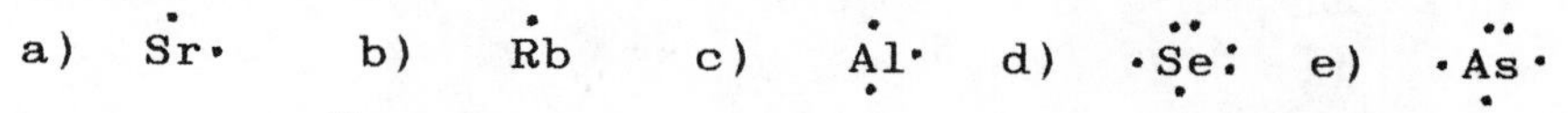 a) Sr·   b) Rb·   c) Al·   d) ·Se: e) ·As·

2. 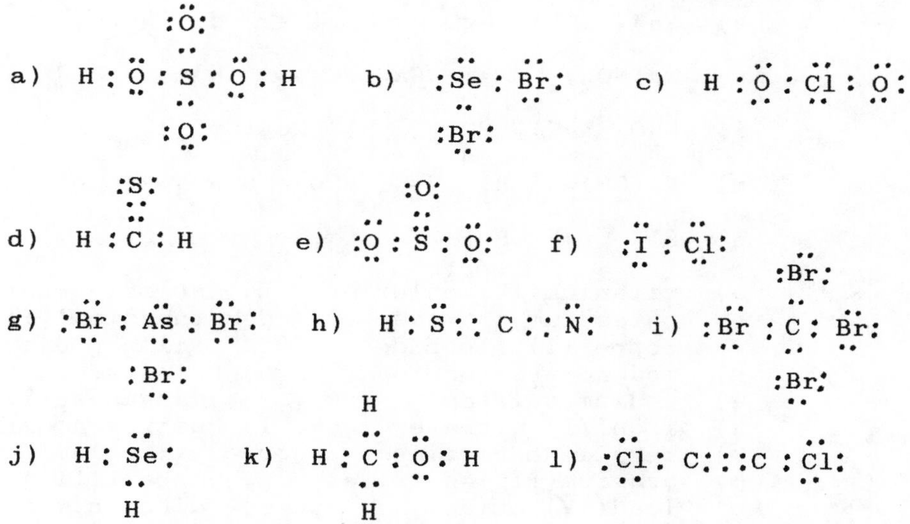 a) H : O : S : O : H   b) :Se : Br: c) H : O : Cl : O :

    d) H : C : H   e) :O : S : O :   f) :I : Cl:

    g) :Br : As : Br: h) H : S :: C :: N: i) :Br : C : Br:

    j) H : Se: k) H : C : O : H   l) :Cl: C ::: C : Cl:

3. a) polar   b) nonpolar   c) polar   d) nonpolar
    e) nonpolar   f) polar   g) polar   h) nonpolar

4. a) Ca^{2+} and N^{3-} make Ca$_3$N$_2$; calcium nitride

    b) Fr$^+$ and P^{3-} make Fr$_3$P; francium phosphide

    c) Na$^+$ and I$^-$ make NaI; sodium iodide

    d) Li$^+$ and O^{2-} make Li$_2$O; lithium oxide

    e) Mg^{2+} and Cl$^-$ make MgCl$_2$; magnesium chloride

    f) Rb$^+$ and N^{3-} make Rb$_3$N; rubidium nitride

    g) Al^{3+} and O^{2-} make Al$_2$O$_3$; aluminum oxide

    h) Ba^{2+} and Br$^-$ make BaBr$_2$; barium bromide

5. a) C (EN = 2.5), S (EN = 2.5); covalent
    b) P (EN = 2.1), I (EN = 2.5); covalent

43

c) Na (EN = 0.9), O (EN = 3.5); ionic
d) S (EN = 2.5), O (EN = 3.5); covalent
e) Cs (EN = 0.7), S (EN = 2.5); ionic
f) As (EN = 2.0), H (EN = 2.1); covalent
g) Li (EN = 1.0), Cl (EN = 3.0); ionic
h) F (EN = 4.0), I (EN = 2.5); covalent

6. a) 3+   b) 3+   c) 2+   d) 7+   e) 4+   f) 3+
   g) 4+   h) 5+   i) 4+   j) 2+   k) 6+   l) 2+

7. a) FeF_3 b) HgO   c) IF_5 d) Na_3P e) LiCl

   f) $Mg(NO_2)_2$ g) $Ba_3(PO_4)_2$ h) $Sn(C_2H_3O_2)_4$

   i) N_2O_5 j) PCl_3 k) $Co_2(CrO_4)_3$ l) $SrCO_3$

   m) $Ni(OH)_3$ n) V_2S_5 o) $U(PO_4)_2$ p) $Bi_2(SO_4)_3$

   q) KI   r) CBr_4

8. a) titanium(IV) chloride   b) sulfur hexafluoride
   c) potassium nitride   d) tungsten(VI) oxide
   e) copper(I) fluoride   f) silicon dioxide
   g) iodine trifluoride   h) dinitrogen tetroxide
   i) sodium sulfide   j) manganese(IV) sulfite
   k) iron(II) nitrate   l) nitrogen oxide
   m) beryllium nitride   n) strontium hydroxide
   o) uranium(III) nitrite   p) cobalt(III) arsenate
   q) lead(IV) oxide   r) diboron trioxide

Multiple Choice

1. c   2. a   3. d   4. d
5. c (should be $Ca_3(PO_4)_2$)
6. c (a should be calcium bromide,
      b should be tetraphosphorus pentoxide,
      d should be manganese(IV) oxide)
7. c   8. d   9. a   10. b

5

CHEMICAL REACTIONS

CHAPTER OVERVIEW

This chapter deals with writing chemical equations and using coefficients to balance them. Many inorganic reactions can be classified into the four main categories which are presented. The activity series is mentioned in conjunction with single replacement reactions. Solubilities are correlated with the precipitation subcategory of double replacement reactions. Oxidation-reduction reactions are identified and discussed. Calculations are done which relate the masses of the components in the balanced equation

TOPIC SUMMARIES AND EXAMPLES

Introduction

Chemical reactions are all around us and are present in all of our life functions. This section gives many examples of reactions with which you are familiar.

Writing Formula Equations for Word Equations; The Chemist's Shorthand

Just as symbols are a shorthand for elements and formulas are concise ways of representing compounds, chemical reactions and physical changes can be shown in compact **chemical equations**. By convention, **reactants** (the species which change) are shown on the left hand side and **products** (the results of the reaction) are on the right. An **arrow** (-->) meaning yields or produces separates the two sides. A proper chemical equation has correct formulas for all species present. You will need to review your naming rules and the charges of ions presented in Chapter 4. Sometimes the physical states of the species are indicated by abbreviations in parentheses. The commonly used notations are: (s)-solid, (l)-liquid, (g)-gas and (aq)-aqueous solution (dissolved in water to make a homogeneous mixture). Reaction conditions are sometimes shown above the arrow.

A **balanced equation** not only shows what reacts and what is produced, but also indicates the proper amounts of each in accordance with the law of conservation of mass. Equations are balanced by using coefficients which multiply the whole species. Formulas or symbols, once correctly written, are not to be modified in any manner. As in formulas for compounds, a coefficient of "1" is considered to be understood and is omitted in the final balanced equation.

Example 5.1

Write the balanced equations for the following word equations.

a) Liquid carbon disulfide reacts with oxygen gas to produce the gases carbon dioxide and sulfur dioxide.
b) When solid aluminum metal is added to an aqueous solution of copper(II) nitrate, solid copper metal and aqueous aluminum nitrate result.
c) Solid sodium reacts with liquid bromine to produce solid sodium bromide.

Solution

a) The balanced equation is:

$$CS_2(l) + 3O_2(g) \longrightarrow CO_2(g) + 2SO_2(g)$$

There are 1 C atom, 2 S atoms and 6 O atoms on each side of the equation.

b) The balanced equation is:

$$3Cu(NO_3)_2(aq) + 2Al(s) \longrightarrow 2Al(NO_3)_3(aq) + 3Cu(s)$$

There are 3 Cu atoms, 6 N atoms, 2 Al atoms and 18 O atoms on each side of the equation.

c) The balanced equation is:

$$2Na(s) + Br_2(l) \longrightarrow 2NaBr(s)$$

There are 2 Na atoms and 2 Br atoms on each side of the equation.

Chemical Reactions: A Look at Four Important Types

Combination reactions are those in which one compound is formed from two or more simpler substances such as:

$$CaO(s) + H_2O(l) \longrightarrow Ca(OH)_2(aq)$$

$$2MgO(s) + O_2(g) \longrightarrow 2MgO(s)$$

Decomposition reactions are the reverse of combination reactions. They involve the breakdown of a compound into two or more components such as:

$$(NH_4)_2Cr_2O_7(s) \longrightarrow N_2(g) + Cr_2O_3(s) + 4H_2O(g)$$
$$\text{heat}$$
$$C_6H_{12}O_6(s) \longrightarrow 6C(s) + 6H_2O(g)$$

Single replacement reactions are those in which one free element replaces another element in a compound such as:

$$2Li(s) + 2H_2O(l) \longrightarrow 2LiOH(aq) + H_2(g)$$

$$F_2(g) + 2KBr(aq) \longrightarrow 2KF(aq) + Br_2(aq)$$

The ability of one element to replace another depends on their relative **activities**. Table 5-1 in the text lists an **activity series**. A more active element can replace a less active element below it on the table. Remember to use the characteristic ion charges in the compounds produced.

Example 5.2

Will the following reactions occur? If so, balance the resulting equations.

a) $Mg(s) + Al(NO_3)_3(aq) \longrightarrow$ b) $Cu(s) + SnCl_2(aq) \longrightarrow$

c) $Cl_2(g) + CaI_2(aq) \longrightarrow$

Solution

a) This occurs because Mg is above Al.

   $$3Mg(s) + 2Al(NO_3)_3(aq) \longrightarrow 3Mg(NO_3)_2(aq) + 2Al(s)$$

b) This does not occur since Sn is above Cu.

c) This occurs since Cl is above I.

   $$Cl_2(g) + CaI_2(aq) \longrightarrow CaCl_2(aq) + I_2(aq)$$

Double Replacement reactions are reactions in which two compounds exchange ions such as:

$$2NaOH(aq) + H_2SO_4(aq) \longrightarrow Na_2SO_4(aq) + 2H_2O(l)$$

$$Ba(OH)_2(aq) + Na_2SO_4(aq) \longrightarrow BaSO_4(s) + 2NaOH(aq)$$

Double replacement reactions are subclassified into neutralization and precipitation reactions. In a **neutralization** reaction, an acid (H^+ donor) reacts with a base (OH^- donor) to yield a salt and water. In a **precipitation** reaction, an **insoluble** compound (a precipitate) is formed when solutions of aqueous compounds are mixed together. Solubility tables are found in Table 5-2 and Appendix C in the text. Additionally, double replacement reactions can occur if a gaseous product is formed. An ion exchange which does not cause formation of water, a gas or a precipitate is not considered to be a double replacement reaction since nothing new is formed; the ions are simply written as if they were rearranged.

Example 5.3

Write balanced equations for the following reactions if they occur.

   a) $H_2SO_4(aq) + KOH(aq) \longrightarrow$

   b) $KBr(aq) + AgNO_3(aq) \longrightarrow$

   c) $KCl(aq) + Ca(C_2H_3O_2)_2(aq) \longrightarrow$

   d) $H_3PO_4(aq) + LiOH(aq) \longrightarrow$

   e) $Ca(CN)_2(aq) + HCl(aq) \longrightarrow$ (hint-HCN is a gas)

Solution

   a) $H_2SO_4(aq) + 2KOH(aq) \longrightarrow K_2SO_4(aq) + 2H_2O(l)$

   b) $KBr(aq) + AgNO_3(aq) \longrightarrow AgBr(s) + KNO_3(aq)$

   c)   No reaction--$KC_2H_3O_2$ and $CaCl_2$ are both soluble.

   d) $H_3PO_4(aq) + 3LiOH \longrightarrow Li_3PO_4(aq) + 3H_2O(l)$

   e) $Ca(CN)_2(aq) + 2HCl(aq) \longrightarrow CaCl_2(aq) + 2HCN(g)$

Oxidation-Reduction Reactions: Another Way to Classify REactions

   Oxidation-reduction reactions, also known as redox

48

reactions are electron transfer reactions. **Oxidation** used to be defined as the reaction of a substance with oxygen. It is now defined as a loss of electrons by an element, resulting in a higher oxidation number. **Reduction** is the gain of electrons by an element, resulting in a lower oxidation number. Oxidation must always be coupled with reduction. In other words, something must give up electrons so that another species can gain them. The **oxidizing agent** is or contains the element which accepts electrons from the substance being oxidized and is itself reduced. The **reducing agent** gives up electrons to cause another species to be reduced and is or contains the element which is oxidized in the process. All single replacement reactions and most combination and decomposition reactions are redox reactions. Double replacement reactions are never oxidation-reduction reactions. Examples of commmonly encountered oxidizing and reducing agents are given in the text. Review the rules for assigning oxidation numbers in Chapter 4. Note that coefficients have no effect on the oxidation numbers of the elements within a compound.

Example 5.4

For each of the following reactions: 1) assign oxidation numbers to the elements; 2) tell what element is oxidized; 3) tell what element is reduced; 4) tell what the oxidizing agent is; and 5) tell what the reducing agent is.

a) $Zn(s) + CuCl_2(aq) \rightarrow ZnCl_2(aq) + Cu(s)$

b) $2CO(g) + O_2(g) \rightarrow 2CO_2(g)$

c) $2K(s) + 2H_2O(l) \rightarrow 2KOH(aq) + H_2(g)$

d) $2Ag(s) + S(s) \rightarrow Ag_2S(s)$

e) $3Mg(s) + N_2(g) \rightarrow Mg_3N_2(s)$

Solution

a)
$$\overset{0}{Zn}(s) + \overset{+2 \ -1}{CuCl_2}(aq) \rightarrow \overset{+2 \ -1}{ZnCl_2}(aq) + \overset{0}{Cu}(s)$$

Zn has an increase in oxidation number; it is oxidized and is the reducing agent. Cu has a decrease in oxidation number; it is reduced and $CuCl_2$ is the oxidizing agent.

b)
$$\overset{+2 \ -2}{2CO}(g) + \overset{0}{O_2}(g) \rightarrow \overset{+4 \ -2}{2CO_2}(g)$$

C has an increase in oxidation number; it is oxidized and CO is the reducing agent. O has a decrease in

49

oxidation number; it is reduced and O_2 is the oxidizing agent.

$$
\begin{array}{cccc}
0 & +1-2 & +1-2+1 & 0 \\
\end{array}
$$

c) 2K(s) + 2H$_2$O(l) --> 2K OH(aq) + H$_2$(g)

K has an increase in oxidation number; it is oxidized and is the reducing agent. H has a decrease in oxidation number; it is reduced and H$_2$O is the oxidizing agent.

$$
\begin{array}{ccc}
0 & 0 & +1 \; -2 \\
\end{array}
$$

d) 2Ag(s) + S(s) --> Ag$_2$S(s)

Ag has an increase in oxidation number; it is oxidized and is the reducing agent. S has a decrease in oxidation number; it is reduced and is the oxidizing agent.

$$
\begin{array}{ccc}
0 & 0 & +2 \; -3 \\
\end{array}
$$

e) 3Mg(s) + N$_2$(g) --> Mg$_3$N$_2$(s)

Mg has an increase in oxidation number; it is oxidized and is the reducing agent. N has a decrease in oxidation number; it is reduced and N$_2$ is the oxidizing agent.

Chemical Calculations Using the Balanced Equation

A balanced chemical equation gives **relative ratios** of molecules or moles of molecules which combine and are formed in a reaction. Using this information combined with a knowledge of formula weights (Review Chapter 2), we can predict amounts of reactants needed to combine with each other and amounts of products which may be formed.

Example 5.5

Use the balanced equation below to calculate: a) the number of moles of I$_2$ which are needed to react with 7.0 moles of K; and b) the number of moles of KI which can be formed in the process.

$$2K(s) + I_2(s) --> 2KI(s)$$

Solution

a) The balanced equation tells us that 2 moles of K will react with 1 mole of I$_2$.

$$? \text{ mol } I_2 = 7.0 \text{ mol K} \times \frac{1 \text{ mol } I_2}{2 \text{ mol K}} = \textbf{3.5 mol } I_2$$

b) The balanced equation tells that 2 mole of K will produce 2 moles of KI.

$$? \text{ mol KI} = 7.0 \text{ mol K} \times \frac{2 \text{ mol KI}}{2 \text{ mol K}} = \textbf{7.0 mol KI}$$

Example 5.6

Use the previous balanced equation to calculate: a) the number of grams of K; and b) the number of grams of I_2 which are needed to prepare 332 grams of KI.

Solution

The formula weights of K, I_2 and KI are 39 g/mol, 254 g/mol and 166 g/mol, respectively. First, we need to calculate the number of moles of KI to be made.

$$? \text{ mol KI} = 332 \text{ g KI} \times \frac{1 \text{ mol KI}}{166 \text{ g KI}} = 2.00 \text{ mol KI}$$

We can then find the amounts of K and I_2 needed to produce this number of moles of KI.

a) $$? \text{ mol K} = 2.00 \text{ mol KI} \times \frac{2 \text{ mol K}}{2 \text{ mol KI}} = 2.00 \text{ mol K}$$

$$? \text{ g K} = 2.00 \text{ mol K} \times \frac{39 \text{ g K}}{1 \text{ mol K}} = \textbf{78.0 g K}$$

b) $$? \text{ mol } I_2 = 2.00 \text{ mol KI} \times \frac{1 \text{ mol } I_2}{2 \text{ mol KI}} = 1.00 \text{ mol } I_2$$

$$? \text{ g } I_2 = 1.00 \text{ mol } I_2 \times \frac{254 \text{ g } I_2}{1 \text{ mol } I_2} = \textbf{254 g } I_2$$

Note that the 78 g K + 254 g I_2 = 332 g KI. Mass is conserved when both the coefficients and the formula weights of the involved species are considered. Similar calculations can be performed for any pair of reactants and/or products when the balanced equation and formula weights are consulted.

SELF-TEST

Fill in the Blanks

1. The substance which gives up electrons is the _____ agent and is itself _____.

2. Equations may be balanced by putting _____ in front of formulas or symbols.

3. The reaction of an acid and a base yields a _____ and _____ and is called a _____ reaction as well as a double replacement reaction.

4. A _____ is an insoluble substance formed when two aqueous solutions are mixed.

5. A single replacement reaction will occur when the _____ of a free element is greater than that of an element in a compound.

True or False

1. Calcium carbonate is soluble in water.

2. Potassium is more active than zinc.

3. $Mg(OH)_2$ can be used to neutralize an acid.

4. In the equation $2Fe(s) + 3S(s) \longrightarrow 2Fe_2S_3(s)$, the iron is the reducing agent.

5. When $Mg(NO_3)_2(aq)$ is reacted with $KCl(aq)$, a precipitate will form.

6. By the equation $2Al(s) + 3S(s) \longrightarrow Al_2S_3(s)$, 27 grams of Al will react with 48 grams of S.

7. $CH_3OH(l) + 2O_2(g) \longrightarrow CO_2(g) + 2H_2O(g)$ is a balanced equation.

8. The sum of the coefficients when this reaction is balanced is 11. $P_4O_{10}(g) + H_2O(l) \longrightarrow H_3PO_4(aq)$

9. No reaction occurs when solid Ca is added to an aqueous solution of KCl .

Short Answer

1. Write and balance the chemical equations for the

52

following word equations.

a) When aqueous potassium chloride is added to aqueous lead(II) nitrate, solid lead(II) chloride and a solution of aqueous potassium nitrate are produced.

b) The reaction of aqueous barium cyanide (cyanide is CN^-) with aqueous hydrogen chloride (more commonly known as hydrochloric acid) produces aqueous barium chloride and hydrogen cyanide gas.

c) The reaction of solid iron with gaseous chlorine produces solid iron(III) chloride.

d) The reaction of solid cesium with liquid water creates a solution of aqueous cesium hydroxide and releases hydrogen gas.

e) Solid potassium nitrate, when heated, yields solid potassium nitrite and oxygen gas.

2. What type of reaction is each reaction in question 1?

3. Balance the following equations.

a) $N_2O_5(s) + H_2O(l) \rightarrow HNO_3(aq)$

b) $H_2O_2(l) \rightarrow H_2O(l) + O_2(g)$

c) $CaBr_2(aq) + Na_2SO_4(aq) \rightarrow CaSO_4(s) + NaBr(aq)$

d) $NiI_3(aq) + Hg(l) \rightarrow Ni(s) + HgI_2(s)$

e) $Al_2O_3(s) + H_2O(l) \rightarrow Al(OH)_3(s)$

f) $Al(NO_3)_3(aq) + CsOH(aq) \rightarrow$
$CsNO_3(aq) + Al(OH)_3(s)$

g) $KOH(aq) + H_3AsO_4(aq) \rightarrow K_3AsO_4(aq) + H_2O(l)$

h) $NaHCO_3(s) \xrightarrow{heat} Na_2CO_3(s) + H_2O(g) + CO_2(g)$

i) $CrCl_3(aq) + Mg(s) \rightarrow MgCl_2(aq) + Cr(s)$

j) $Mg_3N_3(s) + H_2O(l) \rightarrow Mg(OH)_2(s) + NH_3(g)$

4. Into which reaction category does each of the reactions in question 3 fall?

5. Predict the products and balance the following single replacement reactions if they will occur.

a) $Mg(s) + AuCl_3(aq) \rightarrow$

b) $Ag(s) + Ca(NO_3)_2(aq) \rightarrow$

c) $AlBr_3(aq) + Cl_2(aq) \rightarrow$

d) $Cr(C_2H_3O_2)_3(aq) + Al(s) \rightarrow$

e) $Zn\ Cl_2(aq) + H_2(g) \rightarrow$

f) $Ca(s) + H_2O(l) \rightarrow$

6. Predict the products and balance the following double replacement reactions if they will occur.

a) $CdSO_4(aq) + H_2S(aq) \rightarrow$

b) $Ba(OH)_2(aq) + HNO_3(aq) \rightarrow$

c) $NaCl(aq) + (NH_4)_2S(aq) \rightarrow$

d) $SrI_2(aq) + AgNO_3(aq) \rightarrow$

e) $Mg(ClO_3)_2(aq) + LiOH(aq) \rightarrow$

f) $LI_3PO_4(aq) + Ga(NO_3)_3(aq) \rightarrow$

7. Balance the following reactions then assign oxidation numbers to decide if the following reactions are redox reactions. For each redox reaction, tell which element is oxidized and which element is reduced.

a) $Al(s) + HCl(aq) \rightarrow AlCl_3(aq) + H_2(g)$

b) $Br_2(l) + KI(aq) \rightarrow KBr(aq) + I_2(aq)$

c) $Cd(s) + N_2(g) \rightarrow Cd_3N_2(s)$

d) $H_2S(g) + O_2(g) \rightarrow H_2O(g) + SO_2(g)$

e) $CaO(s) + CO_2(g) \rightarrow CaCO_3(s)$

f) $Rb(s) + H_2O(l) \rightarrow RbOH(aq) + H_2(g)$

8. What is the oxidizing agent and what is the reducing agent in each redox reaction in question 7?

9. Balance the equation below and use it to answer the following questions.

$P_4(s) + H_2(g) \rightarrow PH_3(g)$

a) How many moles of H_2 are needed to react with 1.6

moles of P_4?

b) How many moles of PH_3 can be made from 1.6 moles of P_4?

c) How many moles of P_4 are needed to make 8.0 moles of PH_3?

d) How many moles of H_2 must react to make 8.0 moles of PH_3?

e) How many grams of P_4 are needed to make 17.0 grams of PH_3?

f) How many grams of H_2 are needed to make 17 grams of PH_3?

10. Balance the following equation and use it to answer the following questions.

$$Sn(s) + Br_2(l) \longrightarrow SnBr_4(s)$$

a) How many grams of $SnBr_4$ can be made from 10 grams of Sn?

b) How many grams of Br_2 are needed to react with 10 grams of Sn?

c) How many grams of Sn are needed to make 100 grams of $SnBr_4$?

d) How many grams of Br_2 are needed to make 100 grams of $SnBr_4$?

11. Balance the following equation and use it to answer the following questions.

$$NaOH(aq) + H_2SO_4(aq) \longrightarrow Na_2SO_4(aq) + H_2O(l)$$

a) How many grams of H_2SO_4 are needed to react with 20 grams of NaOH?

b) How many grams of Na_2SO_4 can be made from 20 grams of NaOH?

c) How many grams of H_2O can be made from 20 grams of NaOH?

d) How many grams of NaOH are needed to make 7.1 grams of Na_2SO_4?

e) How many grams of H_2SO_4 are needed to make 7.1 grams of Na_2SO_4?

f) How many grams of H_2O are made when 7.1 grams of Na_2SO_4 are also produced?

Multiple Choice

1. When $Ca(OH)_2(aq)$ and $H_2SO_4(aq)$ are mixed the precipitate formed is:

a) Ca_2SO_4 b) $CaSO_4$ c) H_2O
d) no precipitate is formed

2. What is the sum of the coefficients needed to balance the reaction below?

$(NH_4)_2SO_4 + KOH \rightarrow K_2SO_4 + NH_3 + H_2O$

a) 11   b) 9   c) 6   d) 8

3. Which element is oxidized in the following reaction?

$2 PbCl_2(aq) + Sn(s) \rightarrow SnCl_4(aq) + 2 Pb(s)$

a) Pb   b) Cl   c) Sn   d) not a redox reaction

4. In question 3, what is the reducing agent?

a) $SnCl_4$ b) Pb   c) $PbCl_2$
d) not a redox reaction

5. Which of the following combinations will not result in a precipitate being formed?

a) $Ca(OH)_2(aq)$, $H_2SO_4(aq)$ b) $Na_2SO_3(aq)$, $AuCl_3(aq)$
c) $NaBr(aq)$, $Ca(NO_3)_2(aq)$
d) $AlCl_3(aq)$, $(NH_4)_3PO_4(aq)$

6. Which combination in question 5 would yield a neutralization reaction?

7. $KClO_3(s) \rightarrow KCl(s) + O_2(g)$

What is the sum of the coefficients when this equation is balanced?

a) 3   b) 5   c) 7   d) 9

Use the balanced equation below to answer questions 8-12.

$4FeS(s) + 7O_2(g) \rightarrow 2Fe_2O_3(s) + 4SO_2(g)$

8. How many moles of SO_2 can be formed from 3.5 moles of O_2?

a) 7.0 mol SO_2 b) 6.1 mol SO_2 c) 2.0 mol SO_2
d) 1.0 mol SO_2

9. How many grams of O_2 are needed to react with 22 grams of FeS?

a) 14 g O_2 b) 38 g O_2 c) 8.0 g O_2 d) 7 g O_2

10. How many grams of Fe_2O_3 can be made from 11 g FeS?

   a) 20 g Fe_2O_3 b) 10 g Fe_2O_3 c) 5.5 g Fe_2O_3
   d) 40 g Fe_2O_3

11. How many grams of SO_2 are made when 10 grams of Fe_2O_3 are also formed?

   a) 5.0 g SO_2 b) 4.0 g SO_2 c) 50 g SO_2
   d) 8.0 g SO_2

12. How many grams of O_2 must react to make 8.0 grams of SO_2?

   a) 7.0 g O_2 b) 3.5 g O_2 c) 14 g O_2
   d) 4.0 g O_2

ANSWERS TO SELF-TEST

Fill in the Blanks

1. reducing; oxidized   2. coefficients
3. salt; water; neutralization   4. precipitate
5. activity

True or False

1. False--see solubility table   2. True   3. True
4. True   5. False--$MgCl_2$ and KNO_3 are both soluble
6. True
7. False--$2CH_3OH + 3O_2 \rightarrow 2CO_2 + 4H_2O$ is the balanced equation)
8. True--$P_4O_{10} + 6H_2O \rightarrow 4H_3PO_4$
9. True--Ca is less active than K

Short Answer

1. a) $2KCl(aq) + Pb(NO_3)_2(aq) \rightarrow PbCl_2(s) + 2KNO_3(aq)$

   b) $Ba(CN)_2(aq) + 2HCl(aq) \rightarrow 2HCN(g) + BaCl_2(aq)$

   c) $2Fe(s) + 3Cl_2(g) \rightarrow 2FeCl_3(s)$

   d) $2Cs(s) + 2H_2O(l) \rightarrow 2CsOH(aq) + H_2(g)$

   e) $2KNO_3(s) \rightarrow 2KNO_2(s) + O_2(g)$

2. a) double replacement   b) double replacement
   c) combination   d) single replacement
   e) decomposition

3. a) $N_2O_5(s) + H_2O(l) \longrightarrow 2HNO_3(aq)$

b) $2H_2O_2(l) \longrightarrow 2H_2O(l) + O_2(g)$

c) $CaBr_2(aq) + Na_2SO_4(aq) \longrightarrow CaSO_4(s) + 2NaBr(aq)$

d) $2NiI_3(aq) + 3Hg(l) \longrightarrow 2Ni(s) + 3HgI_2(s)$

e) $Al_2O_3(s) + 3H_2O(l) \longrightarrow 2Al(OH)_3(s)$

f) $Al(NO_3)_3(aq) + 3CsOH(aq) \longrightarrow$
$3CsNO_3(aq) + Al(OH)_3(s)$

g) $3KOH(aq) + H_3AsO_4(aq) \longrightarrow K_3AsO_4(aq) + 3H_2O(l)$

h) $2NaHCO_3(s) \overset{heat}{\longrightarrow} Na_2CO_3(s) + H_2O(g) + CO_2(g)$

i) $2CrCl_3(aq) + 3Mg(s) \longrightarrow 3MgCl_2(aq) + 2Cr(s)$

j) $Mg_3N_2(s) + 6H_2O(l) \longrightarrow 3Mg(OH)_2(s) + 2NH_3(g)$

4. a) combination   b) decomposition
c) double replacement   d) single replacement
e) combination   f) double replacement
g) double replacement   h) decomposition
i) single replacement   j) double replacement

5. a) $3Mg(s) + 2AuCl_3(aq) \longrightarrow 2Au(s) + 3MgCl_2(aq)$

b) No reaction (Ca is more active than Ag)

c) $2AlBr_3(aq) + 3Cl_2(aq) \longrightarrow 2AlCl_3(aq) + 3Br_2(aq)$

d) $Cr(C_2H_3O_2)_3(aq) + Al(s) \longrightarrow$
$Al(C_2H_3O_2)_3(aq) + Cr(s)$

e) No reaction (Zn is more active than H)

f) $Ca(s) + 2H_2O(l) \longrightarrow Ca(OH)_2(aq) + H_2(g)$

6. a) $CdSO_4(aq) + H_2S(aq) \longrightarrow CdS(s) + H_2SO_4(aq)$

b) $Ba(OH)_2(aq) + 2HNO_3(aq) \longrightarrow$
$Ba(NO_3)_2(aq) + 2 H_2O(l)$

c) No reaction (NH_4Cl and Na_2S are soluble)

d) $SrI_2(aq) + 2 AgNO_3(aq) \longrightarrow Sr(NO_3)_2(aq) + 2AgI(s)$

e) $Mg(ClO_3)_2(aq) + 2LiOH(aq) \longrightarrow$
$2LiClO_3(aq) + Mg(OH)_2(s)$

f) $Li_3PO_4(aq)$ + $Ga(NO_3)_3(aq)$ -->
$GaPO_4(s)$ + $3LiNO_3(aq)$

7. a)   0   +1-1   +3-1   0

      $2Al(s)$ + $6HCl(aq)$ --> $2AlCl_3(aq)$ + $3H_2(g)$
      Al is oxidized and H is reduced

    b)   0   +1-1   +1-1   0

      $Br_2(l)$ + $2KI(aq)$ --> $2KBr(aq)$ + $I_2(aq)$
      Br is reduced and I is oxidized.

    c)   0   0   heat +2 -3

      $3Cd(s)$ + $N_2(g)$ --> $Cd_3N_2(s)$
      Cd is oxidized and N is reduced.

    d)   +1-2   0   +1-2   +4-2

      $2H_2S(g)$ + $3O_2(g)$ --> $2H_2O(g)$ + $2SO_2(g)$
      S is oxidized and O is reduced.

    e)   +2-2   +4-2   +2+4-2

      $CaO(s)$ + $CO_2(g)$ --> $Ca CO_3(s)$
      Oxidation numbers do not change; not a redox
      reaction.

    f)   0   +1-2   +1-2+1   0

      $2Rb(s)$ + $2H_2O(l)$ --> $2Rb OH(aq)$ + $H_2(g)$
      Rb is oxidized and H is reduced.

8. a) Reducing agent-Al;   oxidizing agent-HCl
    b) Oxidizing agent-Br_2; reducing agent-KI.
    c) Reducing agent-Cd;   oxidizing agent-N_2.
    d) Oxidizing agent-O_2;   reducing agent-H_2S.
    e) Not a redox reaction
    f) Reducing agent-Rb;   oxidizing agent H_2O.

9. Balanced: $P_4(s)$ + $6H_2(g)$ --> $4PH_3(g)$

    FW P_4 = 124; FW H_2 = 2.0; FW PH_3 = 34.

    a) 9.6 mol H_2 (1.6 x 6/1)
    b) 6.4 mol PH_3 (1.6 x 4/1)
    c) 2.0 mol P_4 (8.0 x 1/4)
    d) 12 mol H_2 (8.0 x 6/4)
    e) 15.5 g P_4 (17.0 x 1/34 x 1/4 x 124)
    f) 1.5 g H_2 (17.0 x 1/34 x 6/4 x 2.0)

10. Balanced: $Sn(s)$ + $2Br_2(l)$ --> $SnBr_4(s)$

    FW Sn = 119; FW Br_2 = 160; FW $SnBR_4$ = 439.

a)   37 g $SnBr_4$ (10 x 1/119 x 1/1 x 439)
b)   27 g Br_2 (10 x 1/119 x 2/1 x 160)
c)   27.1 g Sn   (100 x 1/439 x 1/1 x 119)
d)   72.9 g Br_2 (100 x 1/439 x 2/1 x 160)

11.   Balanced: $2NaOH(aq) + H_2SO_4(aq) \longrightarrow$
                         $Na_2SO_4(aq) + 2H_2O(l)$

FW NaOH = 40;   FW H_2SO_4 = 98;   FW Na_2SO_4 = 142;
FW H_2O = 18

a)   25 g H_2SO_4 (20 x 1/40 x1/2 x 98)
b)   36 g Na_2SO_4 (20 x 1/40 x 1/2 x 142)
c)   9.0 g H_2O (20 x 1/40 x 2/2 x 18)
d)   4.0 g NaOH   (7.1 x 1/142 x 2/1 x 40)
e)   4.9 g H_2SO_4 (7.1 x 1/142 x 1/1 x 98)
f)   1.8 g H_2O (7.1 x 1/142 x 2/1 x 18)

Multiple Choice

1.   b
2.   d   (Balanced is $(NH_4)_2SO_4 + 2KOH \longrightarrow$
             $K_2SO_4 + 2NH_3 + 2H_2O$)
3.   c   (Sn changes from 0 to +4 in oxidation number)
4.   c   (Pb changes from +2 to 0 in oxidation number)
5.   c   (both $NaNO_3$ and $CaBr_2$ are soluble)
6.   a   ($Ca(OH)_2$ is a base and H_2SO_4 is an acid)
7.   c   (Balanced is $2KClO_3 \longrightarrow 2KCl + 3O_2$)

For 8-12, FW FeS = 88;   FW O_2 = 32;   FW Fe_2O_3 = 160;
FW SO_2 = 64

8.   c   (3.5 x 7/4)
9.   a   (22 x 1/88 x 7/4 x 32)
10.   b   (11 x 1/88 x 2/4 x 160)
11.   d   (10 x 1/160 x 4/2 x 64)
12.   a   (8.0 x 1/64 x 7/4 x 32)$lO_3 \longrightarrow$ 2 KCl + 3 O_2)

6

THE ENERGY STORY

CHAPTER OVERVIEW

This chapter discusses energy and its various units and mentions some means of energy production. The concept of enthalpy is introduced and calculations are done to find the heats associated with various reactions. The relationship of power to energy is presented. The importance of solar energy to this planet is stressed. Past energy use and future possibilities are surveyed. Entropy is defined and the second law of thermodynamics is used to explain why, even though energy is conserved, it cannot be used again and again.

TOPIC SUMMARIES AND EXAMPLES

Introduction

Energy is seen as being the basis for operation of the world and for life itself. A scenario mentions some possible future fuels and means by which energy usage to accomplish the same tasks might be decreased.

Matter and Energy

The importance of matter is reviewed and it is mentioned that, as well as mass, every substance has associated with it a **chemical potential energy**. This energy depends upon the type of matter from which the substance is made and the bonds which hold it together. When the substance or substances of interest (called the **system**) reacts, bonds are broken and/or made. Breaking bonds requires energy, while making them usually releases energy. Each reaction has a particular net loss or gain of energy depending on the number and kinds of bonds involved. This change in energy is called the **heat of the reaction**. If heat is given off, as when a log burns, the process is said to be **exothermic**. If heat is absorbed, as is needed to boil water, the process is said to be **endothermic**.

The Law of Conservation of Mass and Energy

Just as mass is conserved in ordinary chemical reactions or physical changes, energy is neither created nor destroyed in these processes, but just changed from one form to another. More sensitive measurements have shown that mass can be turned into energy and vice versa. The **law of conservation of mass and energy** states that the sum total of mass and energy in the universe is constant.

Energy Units

Energy exists in many forms such as mechanical, heat, electrical, and chemical energies. Although they seem quite different, they can all be expressed in terms of their capacity to do work. Energy is measured as it is transferred between two systems. Heat, for example, always flows from a hotter object to a colder object until they come to the same final intermediate temperature. **Temperature** is a measure of the intensity of heat and is discussed in Appendix A. Conversions between the Fahrenheit temperature scale and the Celsius temperature scale are also presented there.

In the SI system, the **joule** (abbrieviated **J** and equal to 1 $kg-m^2/s^2$) is the unit used. The **calorie** which is equal to 4.184 joules, is still widely employed. The calorie (cal) was originally defined as the amount of energy required to change the temperature of one gram of water by 1^oC. A "food" **calorie**, often written as a **Calorie**, is in fact a kilocalorie, equal to 1000 calories. Caloric value of food is computed by the amount of heat that it releases when burned with oxygen. Bodies utilize food for energy by reacting it with oxygen as well, although in a much more complex, stepwise fashion. The temperature change which can be caused by a given amount of energy depends upon the mass of the object and its unique **specific heat capacity**. For water, the specific heat capacity is 1 $cal/g-^oC$.

Example 6.1

How many calories and how many joules are needed to change the temperature of 40 g of water rom 25.0^oC to 30.0^oC?

Solution

The temperature change is $30.0^oC - 25.0^oC = 5.0^oC$.

$$? \text{ calories} = 40 \text{ g} \times \frac{1 \text{ cal}}{\text{g-}^{\circ}\text{C}} \times 5.0^{\circ}\text{C} = \textbf{200 cal}$$

$$? \text{ joules} = 200 \text{ cal} \times \frac{4.184 \text{ J}}{1 \text{ cal}} = \textbf{840 J}$$

Chemical Reactions and Heat Content

Every substance has a certain **heat content** called **enthalpy (H)** associated with it. Enthalpies cannot be absolutely measured; however, we can measure heat changes which result from certain processes by monitoring temperature. These heat changes constitute the **enthalpy of reaction (ΔH)**. Many reactions have been experimentally performed and their ΔH values have been calculated and tabulated. Widely used is **heat of formation (ΔH_f)** data, a small sampling of which appears in Table 6-3 in the text. A heat or enthalpy of formation is the amount of heat absorbed or released in a standard formation reaction in which one mole of the substance of interest is formed from its constituent elements in their standard states. Exothermic reactions are assigned negative values of ΔH_f to indicate that energy has been released. Endothermic reactions are assigned positive values of ΔH_f to indicate that heat has been absorbed.

Tabulated values of ΔH_f can be used to calculate the enthalpy change of many other processes using the formula for heat of a reaction ($\Delta H_{reaction}$):

$$\Delta H_{reaction} = \Delta H_f(\text{products}) - \Delta H_f(\text{reactants})$$

Example 6.2

Use Table 6-3 in the text to calculate the enthalpy of reaction for:

$$2H_2S(g) + 3O_2(g) \longrightarrow 2SO_2(g) + 2H_2O(g)$$

Solution

The heat of formation values are:

$H_2S(g) = -4.8$ kcal/mol; $O_2(g) = 0$ kcal/mol;
$SO_2(g) = -70.96$ kcal/mol and $H_2O(g) = -57.8$ kcal/mol

Note that the state and formula are important to the ΔH_f value. SO_2 has a different value from SO_3 and $H_2O(g)$ has a different value from $H_2O(l)$. Additionally, the values are presented in kcal/mol and must be multiplied by the number of moles of each substance present in the balanced equation.

$$\Delta H_{reaction} = [2\ mol\ SO_2(g)][-70.96\ kcal/mol]$$
$$+ [2\ mol\ H_2O(g)][-57.8\ kcal/mol]$$
$$- [2\ mol\ H_2S(g)][-4.8\ kcal/mol]$$
$$- [3\ mol\ O_2(g)][0\ kcal/mol]$$

$$\Delta H_{reaction} = -141.92 - 115.6 + 9.6 = \textbf{-247.9 kcal}\ (exothermic)$$

The energy of the reaction also depends upon the mass and, thus, the numbers of moles of material used.

Example 6.3

How much energy is released when 17 grams of H_2S (g) is reacted with sufficient O_2 for the equation in Example 6.2?

Solution

According to the equation, when 2 moles of H_2S react, 247.9 kcal are released. We need to find the number of moles of H_2S. It has a formula weight of 34 g/mol.

$$?\ mol\ H_2S = 17\ g \times \frac{1\ mol\ H_2S}{34\ g\ H_2S} = 0.50\ mol\ H_2S$$

$$?\ kcal = 0.50\ mol\ H_2S \times \frac{247.9\ kcal}{2\ mol\ H_2S} = \textbf{62 kcal}$$

Power and Energy

Power is the rate of energy consumption or production. We most often hear of energy reported in **watts**, where a watt is 1 joule/second. Electricity is sold in the form of kilowatt-hours. Power x time = energy.

Example 6.4

How much energy in joules is used to burn a 100 watt bulb for one hour?

Solution

100 watts is 100 joule/sec.

$$?\ joules = 1\ hour \times \frac{60\ min}{1\ hour} \times \frac{60\ sec}{1\ min} \times \frac{100\ J}{1\ sec} = \textbf{360,000 J}$$

Solar Energy: Our Free Energy Source

The earth's internal heat (geothermal energy), tidal forces and nuclear fuel together provide less than one percent of the energy used on this planet. All other energy is directly attributable to the sun. It heats the earth and powers the **water cycle**. Heating causes the water to evaporate, changing the sun's radiant energy into potential energy. As rains fall, they provide kinetic energy for **hydropower**. Different heat capacities of land and water give rise to winds which can also be used as an energy source. The sun also powers the life processes of **photosynthesis** by which plants grow. These plants provide energy as they are eaten. Additionally, fossil fuels (so called because they bear or are found in rock forms bearing fossil remains) such as coal, oil and natural gas, are the result of ancient photosynthesis and the subsequent death and decay of plants and the animals that ate them.

Early Uses of Energy

Man progressed from muscle power to animal power. These depended on foods and, thus, the sun as fuel. Fire was used as a technology for cooking food and for warmth. The **water wheel**, and the **windmill** gave untiring energy sources, again using the sun to replenish their fuel. Combustion or burning reactions lead to the **steam engine** which enabled power production on a large scale to be located far from water or dependable wind sources. The **internal combustion engine** enabled faster forms of tranportation and power generation. **Nuclear fission** currently provides a large amount of the world's electric power. Humankind has shifted from directly sun dependent renewable resources to energy from nonrenewable resources.   At the same time, the demand for energy for many purposes has increased.

The Age of Fossil Fuels: It Will End Soon

Coal is thought to be the decay product of ancient plants from swamps and bogs. Coal is created in a process where soft **peat** is formed first. The peat is gradually converted to **lignite** which is in turn converted to **bituminous coal** and finally to hard **anthracite coal**. All stages can be found in the world and are used as fuel; however, anthracite, which takes on the order of 600,000,000 years to produce has the highest energy per gram since it contains the highest concentration of carbon. Coal is not considered to be a renewable resource, because, although supplies are being gradually replenished, the replacement rate is far lower than the current rate of consumption.

Oil and natural gas are speculated to be the product of animal decay since similarities in structure to those of animal fats have been found. Oil and gas supplies are being replenished, but the renewal rate is dwarfed by the magnitude of the consumption, so they are being seriously depleted.

If Energy Can Neither Be Created Nor Destroyed, Can It Be Recycled?

The **first law of thermodynamics** is the law of conservation of energy. It indicates that energy must come from somewhere else. If energy can't be destroyed, it seems plausible that it might be infinitely reusable. However, the **second law of thermodynamics** states that all processes lead to an increase of randomness or disorder called **entropy**. This means that each time a process occurs, a certain percentage of the energy is converted to waste heat which increases molecular motion and randomness instead of turning into useful work. This **low quality** energy cannot be collected and reused. We are said to be running out of energy because **high quality** energy sources such as natural gas, coal and nuclear fuel are being turned into waste heat along with useful work. Since these sources are primarily nonrenewable, the amount of useful energy is steadily decreasing.

Future Energy Sources

Thermal energy of oceans, **oil shale**, **direct solar power** for either heating or to run **photovoltaic cells** for electric generation, and **nuclear fusion** are mentioned as future options for our energy dependence. Some of these are currently being used to a small degree, as are **geothermal wells**, **tidal power** and new emphasis on **wind power**. The latter are of particular interest since they are nonpolluting, renewable resources.

SELF-TEST

Consult Table 6-3 in the text for the proper enthalpy of formation values.

Fill in the Blanks

1. _____ is the rate of production or consumption of energy and is often measured in _____.

2. A calorie is equal to 4.184 _____.

3. The term used for heat of a reaction is _____.

4. The measure of randomness or disorder is _____.

5. Processes which require heat to occur are _____.

6. _____ energy is responsible for most of the earth's energy supply.

7. Water held behind a dam has _____ energy, while water spilling over the top of the dam has _____ energy.

8. A _____ resource is one which is being used faster than it can be replenished.

9. The _____ law of thermodynamics implies that energy cannot be endlessly recycled because it is converted into _____.

10. Heat is transferred from regions of _____ temperature to those of _____ temperature.

True or False

1. A watt is 1.0 cal/second.

2. Melting butter is an endothermic process.

3. A can of gasoline has potential energy.

4. Enthalpy increases whenever a process occurs.

5. $CaO(s) + CO_2(g) --> CaCO_3(s)$ is a formation reaction.

6. Wood is a renewable resource.

7. A "food calorie" is a kilocalorie.

8. A reaction has a heat of formation equal to +7.93 kcal/mol. The reaction is exothermic.

9. The amount of energy required to change the temperature of an object is dependent upon its mass.

10. A 60 watt light bulb uses about 14 cal/second.

Short Answer

1. How many joules are released when you digest a tablespoon of honey? It has a food value of 65 Calories.

2. How many calories and how many joules are needed to change the temperature of 18 grams of water from

5.0°C to 25.0°C?

3. Calculate the heat of reaction for:

   $H_2O(l) \longrightarrow H_2O(g)$.

   Is this process endothermic or exothermic?

4. Consult question 3 and calculate the energy needed to convert 72 grams of $H_2O(l)$ to $H_2O(g)$ at the boiling point.

5. How many joules are needed to run a 650 watt iron for 15 minutes?

6. Balance the following reaction and calculate the change in enthalpy for the process. The reaction is the combustion of octane, a component of gasoline.

   $C_8H_{18}(l) + O_2(g) \longrightarrow CO_2(g) + H_2O(g)$

7. How much energy is required to decompose 1 mole of $SO_3(g)$ to $SO_2(g)$ and $O_2(g)$?

8. What is the enthalpy of reaction for the decomposition of $CaCO_3(s)$ to $CaO(s)$ and $CO_2(g)$?

Multiple Choice

1. Which process is endothermic?

   a) gasoline burning   b) exploding dynamite
   c) freezing water   d) baking a cake

2. Which is the largest energy unit?

   a) a calorie   b) a Calorie   c) a joule
   d) a kilojoule

3. Which is not a predecessor of anthracite coal?

   a) peat   b) lignite   c) magnetite
   d) dead plants

4. Which is a nonrenewable energy source?

   a) wind power   b) hydropower   c) corn oil
   d) nuclear fuel

5. Which requires the most energy?

    a) heating 100 grams of water from $10^{\circ}C$ to $20^{\circ}C$
    b) heating 10 grams of water from $25^{\circ}C$ to $30^{\circ}C$
    c) heating 50 grams of water from $5.0^{\circ}C$ to $95.0^{\circ}C$
    d) heating 200 grams of water from $25^{\circ}C$ to $50^{\circ}C$

6. The enthalpy of formation of $C_2H_2(g)$ is found to be +54.2 kcal/mol. How many kilocalories are required to form 6.5 grams of C_2H_2 from its elements?

    a) 14 kcal   b) 8.3 kcal   c) 9200 kcal
    d) 352.3 kcal

7. What is the heat of reaction for:

    $2H_2O_2(l) \longrightarrow 2H_2O(l) + O_2(g)$

    a) -23.5 kcal   b) -47.00 kcal   c) -13.0 kcal
    d) -26.0 kcal

8. Which is true of all processes which occur?

    a) energy is released
    b) mass is converted to energy
    c) disorder increases
    d) entropy is used up

ANSWERS TO SELF-TEST

Fill in the Blanks

    1. power; watts   2. joules   3. enthalpy
    4. entropy   5. endothermic   6. solar
    7. potential; kinetic   8. nonrenewable
    9. second; waste heat or disorder
    10. higher; lower

True or False

    1. False--(1.0 joule/second)   2. True--needs heat
    3. True--can be burned for fuel
    4. False--entropy increases
    5. False--$CaCO_3$ must be made from its elements,
            $Ca(s) + C(s) + 3/2 O_2(g) \longrightarrow CaCO_3(s)$
    6. True   7. True
    8. False--positive is endothermic
    9. True   10. True--60 x 1/4.184

69

Short Answer

1. 270,000 J   (65 x 1000 x 4.184)

2. 360 cal   (18 x 1.0 x 20)
   1500 J   (360 x 4.184)
3. 10.5 kcal; endothermic {1(-57.8) - 1(-68.3)}
4. 42 kcal   (72/18 x 10.5)
5. 585,000 J (15 x 60 x 650)
6. $2C_8H_{18}(l) + 25O_2(g) ---> 16CO_2(g) + 18H_2O(g)$

   -2426.1 kcal {16(-94.1) + 18(-57.8) - 2(-59.97)
                  -25(0)}

7. +23.49 kcal [$2SO_3(g)$ --> $2SO_2(g) + O_2(g)$]
   {2(-70.96) + 1(0) - 2(-94.45)
   = 46.98 kcal for 2 mol SO_3; 46.98/2}

8. +42.5 kcal [$CaCO_3(s)$ ---> $CaO_2(s) + O_2(g)$]
   {1(-151.9) + 1(-94.1) - 1(-288.5)}

Multiple Choice

1. d (needs heat)   2. b   3. c
4. d (the other three are related to solar energy)
5. d (5000 cal compared to 1000 for a, 50 for b and
      4500 for c)
6. a (6.5 x 1/26 x 54.2)
7. b [2(-68.3) + 1(0) - 2(-44.8)]
8. c

7

SOLIDS, LIQUIDS, AND GASES: THE THREE STATES OF MATTER

CHAPTER OVERVIEW

This chapter discusses the three states of matter and relates phase changes to potential and kinetic energy. The kinetic molecular theory is presented in general and then specifically applied to gases. The types of crystalline solids and their different properties are introduced. The concept of equilibrium is related to vapor pressure and boiling point. Pressure and its units and the Kelvin temperature scale are discussed in conjunction with a series of gas laws. Practical applications of the gas laws are included.

TOPIC SUMMARIES AND EXAMPLES

Introduction

Table 7-1 gives the contrasting characteristics of solids, liquids and gases. Familiar examples of each of the states are provided. Figure 7-1 shows the states of the elements under normal temperature and pressure conditions. States will change as temperature and pressure change.

The Kinetic Molecular Theory of Matter

The **kinetic molecular theory of matter**, listed in the text, states that matter is composed of tiny particles, constantly in motion. At the same temperature, all particles, no matter what substance or state, have the same average kinetic energy ($1/2 \, mv^2$, where m is particle mass and v is its velocity). As temperature is increased, the kinetic energy of the particles increases, and they move more rapidly. The particles also have potential energy due to attractions and repulsions. When particles collide, energy is conserved, and just transferred from one particle to another. The state of matter is dependent upon the relationship between **cohesive forces**, which are those attractive forces tending to hold the particles together, and **disruptive forces**, which are repulsions and collisions which tend to push particles apart.

Example 7.1

Calculate the kinetic energy in kilojoules of a car with a mass of 908 kilograms (about 2000 pounds) traveling at 24.6 meter/second (about 55 miles per hour).

Solution

Remember that a joule is a $kg\text{-}m^2/s^2$.

$$1/2 \ mv^2 = 1/2 \ (908 \ kg) \times [24.6 \ m/s]^2 = 275,000 \ kg\text{-}m^2/s^2$$

$$? \ kJ = 275,000 \ kg\text{-}m^2/s^2 \times \frac{1 \ J}{1kg\text{-}m^2/s^2} \times \frac{1kJ}{1000 \ J} = \mathbf{275 \ kJ}$$

The Solid State and Crystals

Solids have fixed shape and volume and are incompressible because of strong repulsive forces between the electrons of the already close atoms. Most solids are arranged in orderly geometric patterns called **crystals**. A crystal is defined by particles in a symmetrical structure called a **crystal lattice**. A crystal can also be viewed as being constructed of repeating units called **unit cells**, much as a brick wall has bricks as repeating units. A few solids such as glass are **amorphous solids** with randomly oriented particles.

Atomic, Molecular and Ionic Crystals: A Look at What Binds Them

As a solid is heated, the particles move more and more rapidly until the disruptive kinetic energy overcomes the cohesive forces. At this point, the solid is said to **melt** and the particles are randomly oriented. The temperature at which this occurs is called the **melting point**. At the melting pont, all heat applied goes into the phase change, as opposed to temperature rise, so a liquid has more potential energy than its solid. (Remember that at the melting point, both are at the same temperature and have the same kinetic energy.) Stronger cohesive forces require higher disruptive forces to overcome them. Therefore, a substance with high cohesive forces will have a higher melting point than one with weaker cohesive forces.

Table 7-3 lists the differences between the four types of

72

crystals. **Ionic** crystals are held together by ionic bonds with resultant strong cohesive forces. They are marked by high melting points and only conduct electricity when melted, at which time the ions have the mobility to carry current. All ionic compounds, such as LiCl and Na_2SO_4, exist as ionic crystals. Review Chapter 3 if necessary. **Molecular** crystals (and noble gas atoms) are held together by weak forces between the molecules. These forces increase with increasing polarity of the molecule, but are much smaller than the cohesive forces of ionic compounds. Covalent substances, such as H_2S and O_2, form molecular crystals. They have no ions and never conduct electricity in the pure state. **Atomic** crystals are subdivided into **metallic** and **nonmetallic** crystals. **Nonmetallic** atomic crystals, such as carbon in a diamond and SiC, are held together by huge networks of covalent bonds, with the high cohesive forces giving them very high melting points. They never conduct electricity since they have no ions. **Metallic** crystals are sometimes described as cations in a sea of electrons. The particles are held together by the attraction between the positive centers and the electrons, so melting points tend to be quite high. The electrons are not tightly associated with a particular cation and are free to move. This causes metals to be able to conduct electricity even in the solid state. All metals, such as Cu, Ni and Ag, exist as metallic crystals.

The Liquid State and Evaporation

The liquid state is characterized by random motion of particles and a constant volume, but an indefinite shape. The particles are still quite close together, so the liquids are not very compressible. As particles move in a liquid, surface particles, when involved in collisions, may receive enough kinetic energy to escape the cohesive forces of the other particles and move into the gas state. This process is called **evaporation**. **Evaporative cooling** occurs because, in each case, the particles which escape are those with higher than average kinetic energy. As they leave, the remaining molecules have a slightly lower average kinetic energy and, therefore, a lower temperature.

Equilibrium Vapor Pressure

In an open container, evaporation can continue to occur, with energy lost by evaporating molecules being replenished by transfers of energy from the surroundings, until the liquid has been totally converted to the gas (vapor) state. If the container is covered and contains sufficient liquid, evaporation seems to cease. In actuality, the particles in the gas state are also undergoing collisions and have kinetic energies both higher and lower than the average kinetic

73

energy. If a slow moving gas particle collides with the surface of the liquid, it may be held by the cohesive forces or **condense** back into the liquid. In a closed container at constant temperature, the rate of evaporation will become equal to the rate of condensation and the system is said to be in a state of **dynamic equilibrium**. The pressure which is exerted by the gas particles that are present at that time is said to be the **equilibrium vapor pressure**. This equilibrium vapor pressure depends only upon the cohesive forces of the substance involved and the temperature. High cohesive forces result in low equilibrium vapor pressures. High temperatures cause high equilibrium vapor pressures. Substances which have high equilibrium vapor pressures, such as nail polish remover and rubbing alcohol, evaporate readily and are said to be **volatile**. Some equilibrium vapor pressure values for H_2O are given in Appendix C, Table 5.

The Phenomenon of Boiling

Boiling is described visually as the point at which bubbles formed in a heated liquid break the surface and escape. More precisely, it is defined as the condition which occurs when the equilibrium vapor pressure is equal to the atmospheric pressure. The **boiling point** is the temperature at which this occurs. The **normal boiling point** is the temperature at which the equilibrium vapor pressure equals the pressure of one standard atmosphere (1.00 atm). At the boiling point, as is observed at the melting point, all heat applied goes into the potential energy of the phase change, in this case from from liquid to gas, rather than into the kinetic energy of temperature rise. At the boiling point, the liquid and its vapor have the same kinetic energy, but the vapor has a greater potential energy. Substances with higher cohesive forces have higher boiling points, while those with lower cohesive forces have lower boiling points.

Because boiling depends upon the pressure of the atmosphere as well as the cohesive forces involved, changes in atmospheric pressure change boiling points. Increases in pressure, as in a pressure cooker, require hotter temperatures for the vapor pressure to equal the atmospheric pressure, so the boiling point is raised. Decreasing the atmospheric pressure lowers the boiling point.

The Gaseous State

Gases follow the kinetic molecular theory; however, the particles are widely spaced and have essentially no attractions so they behave quite independently. Gases are very compressible and have neither fixed shape nor volume; instead, they expand to fill any container.

The Pressure, Volume and Temperature of Gases

Pressure is force per unit area. Pressures were originally measured with a barometer invented by Torricelli in which the height of a mercury column was measured. Pressure is reported in atmospheres, as previously mentioned, but also in torr and millimeters of Hg (mm Hg) where:

$$1 \text{ atm} = 760 \text{ torr} = 760 \text{ mm Hg}$$

Example 7.2

Convert 355 torr to atmospheres.

Solution

$$? \text{ atm} = 355 \text{ torr} \times \frac{1 \text{ atm}}{760 \text{ torr}} = 0.467 \text{ atm}$$

You may have also heard of of p.s.i. (pounds per square inch), used for tire pressures, and inches of mercury, used for barometric pressure readings reported on weather forecasts.

Boyle related pressure to volume of a constant mass of a gas at a constant temperature. He found that the volume decreased as the pressure was increased. **Boyle's law** is:

$$P_i V_i = P_f V_f$$

where the **i**s are for initial conditions and the **f**s are for final conditions.

Example 7.3

A gas sample has a volume of 12.8 liters at 1.00 atmosphere. What is the volume if the pressure is increased to 2.00 atmospheres at constant temperature?

Solution

$$P_i = 1.00 \text{ atm} \quad V_i = 12.8 \text{ L} \quad P_f = 2.00 \text{ atm} \quad V_f = \,?$$

Solving Boyle's law for V_f yields:

$$V_f = \frac{P_i V_i}{P_f} = \frac{(1.00 \text{ atm})(12.8 \text{ L})}{(2.00 \text{ atm})} = 6.40 \text{ L}$$

Charles found that increasing the temperature of a given mass of gas at constant pressure caused the volume to increase. **Charles' law** is expressed as:

$$\frac{V_i}{T_i} = \frac{V_f}{T_f}$$

Temperature must be expressed on the **Kelvin (absolute)** temperature scale. It is predicted that the volume of a gas would be 0 at absolute zero (0 K) if the gas did not condense first. Absolute zero is predicted as being at $-273.15\,^{\circ}C$. We will use the conversion that K = 273 + $^{\circ}C$. Kelvin can never be a negative number, because no temperature lower than 0 K is theoretically possible.

Example 7.4

A gas in a balloon has a volume of 250 milliliters at $20.0\,^{\circ}C$. What is its volume at $30.0\,^{\circ}C$ if the pressure is held constant?

Solution

$$V_i = 250 \text{ mL} \quad T_i = 20.0\,^{\circ}C + 273 = 293 \text{ K}$$

$$V_f = ? \quad\quad\quad T_f = 30.0\,^{\circ}C + 273 = 303 \text{ K}$$

Solving Charles' law for V_f yields:

$$V_f = \frac{V_i T_f}{T_i} = \frac{(250 \text{ mL})(303 \text{ K})}{(293 \text{ K})} = 259 \text{ mL}$$

Boyle's Law and Charles' Law can be combined into the **combined gas law** which is:

$$\frac{P_i V_i}{T_i} = \frac{P_f V_f}{T_f}$$

Any constant term can be deleted on both sides yielding the two previous equations. If volume is held constant, a relationship between the pressure and the temperature in kelvins, called **Guy Lussac's law**, is evident.

Example 7.5

A gas with a volume of 6.50 liters at $50\,^{\circ}C$ and 77.0 torr is cooled to $25\,^{\circ}C$ and transferred into a 2.00 liter container. What is the new pressure?

76

Solution

$$P_i = 77.0 \text{ torr} \quad V_i = 6.50 \text{ L} \quad T_i = 50^{\circ}\text{C} + 273 = 323 \text{ K}$$

$$P_f = ? \quad\quad\quad\quad V_f = 2.00 \text{ L} \quad T_f = 25^{\circ}\text{C} + 273 = 298 \text{ K}$$

Solving the combined gas law for P_f yields:

$$P_f = \frac{P_i V_i T_f}{V_f T_i} = \frac{(77.0 \text{ torr})(6.50 \text{ L})(298 \text{ K})}{(2.00 \text{ L})(323 \text{ K})} = \mathbf{231 \text{ torr}}$$

Dalton's law of partial pressures states that, for chemically nonreactive gases, the total pressure exerted by a mixture of gases is the sum of the pressures that each gas would exert if alone in the container (its partial pressure). Additionally, if the total pressure of a gas mixture is known, the pressure due to the individual gases can be computed by multiplying the decimal volume percent by the total pressure.

Example 7.6

In a mixture of gases, the pressure due to O_2 is 1.0 atmosphere and that due to N_2 is 2.5 atmospheres. What is the total pressure?

Solution

The total pressure is the sum of the partial pressures.

? atm = 1.0 atm + 2.5 atm = **3.5 atm**

Tracking Oxygen and Carbon Dioxide

The book traces the partial pressures of oxygen and carbon dioxide in the arteries and the veins. It relates this to Dalton's law of partial pressures and mentions the fact that gases flow from regions of higher pressure to regions of lower pressure. You have also heard this fact in discussions of air flow on weather forecasts.

Henry's Law

Henry's law states that the solubility of a gas in a liquid is directly related to the pressure of that gas in the air above the liquid. **Hyperbaric oxygen therapy** involves putting a patient into a high pressure oxygen atmosphere to dissolve more oxygen in the blood and hasten the healing process. The **bends** in divers is related to the fact that extra nitrogen dissolves in the blood when air is breathed from pressurized air tanks. The nitrogen can bubble out of the blood creating problems if the pressure is released too quickly because of surfacing to normal air pressure at too rapid a pace. Soft drinks are carbonated at high pressures of carbon dioxide. When the cap is removed, the pressure is released and the solution is subjected to the normal carbon dioxide pressure of the air. The extra dissolved carbon dioxide bubbles out in characteristic fizz.

SELF-TEST

Fill in the Blanks

1. Sn is expected to form a _____ crystal.

2. Increasing the temperature, _____ the volume of a gas at constant pressure.

3. A more volatile substance has a higher _____ and lower _____ forces than a less volatile substance.

4. The process of a solid turning into a liquid is _____.

5. At the boiling point, the _____ is equal to the _____.

6. An H_2 molecule has _____ kinetic energy than an O_2 molecule traveling at the same velocity.

7. As atmospheric pressure is increased, boiling point _____.

8. NH_3 should form _____ crystals.

9. _____ forces hold particles together and _____ forces tend to force them apart.

10. The process of a gas turning into a liquid is called _____.

11. _____ is when two opposing processes occur at equal rates.

12. Increasing temperature _____ the average kinetic energy of a substance.

13. Gas flows from regions of _____ pressure to regions of _____ pressure.

14. _____ solids do not have an orderly crystal arrangement.

True or False

1. Equilibrium vapor pressure depends upon atmospheric pressure.

2. $BaCl_2$ should form ionic crystals.

3. Gas solubility increases with pressure of that gas.

4. Solid CO should conduct electricity.

5. The volume of a gas is greater at 550 torr than at 700 torr.

6. A liquid has more potential energy than a solid at the same temperature.

7. The velocity of a CO molecule is less than that of a He molecule at the same temperature.

8. The volume of a gas doubles if the temperature is increased from 250 K to 500 K.

9. $H_2O(l)$ is less compressible than $H_2O(g)$.

10. The normal boiling point is determined at 760 torr.

Short Answer

1. Convert 330 torr to atm.

2. Convert $-30^{o}C$ to K.

3. What kind of crystal is formed by U?

4. What kind of crystal is formed by CCl_4?

5. What is the kinetic energy of a particle with a mass of 0.20 kg traveling at a velocity of 3.0 m/s?

6. A gas with a volume of 12 liters at $15^{o}C$ is heated to $80^{o}C$ at constant pressure. What is the new volume?

7. A sample with a total pressure of 985 torr is 20.0% oxygen by volume. What is the pressure due to the oxygen?

8. A gas has a pressure of 14.0 atmospheres when the volume is 3.00 liters. What is the pressure of the same gas at constant temperature in a 12.0 liter container?

9. A gas with a pressure of 1.00 atmosphere and a volume of 2.00 liters at 0°C is heated to 100°C. If the volume is held constant, what is the new pressure?

10. Which is the higher pressure--1.20 atm or 790 torr?

11. Consult Appendix C, Table 5. What is the boiling point of water at an atmopheric pressure of a) about 50 torr and b) about 355 torr?

12. Consult Appendix C, Table 5. What is the pressure at which water boils at a) 0°C and b) 50°C?

Multiple Choice

1. What is the kinetic energy of a particle with a mass of 7.0 kilograms and a velocity of 10.0 m/s?

    a) 2450 J   b) 1225 J   c) 245 J   d) 350 J

2. Which substance below will conduct electricity in the solid state?

    a) $MgCl_2$ b) Ca   c) $Ti(NO_3)_2$ d) PCl_3

3. Which solid has the lowest melting point?

    a) Cs   b) KBr   c) H_2O d) Fe

4. Which pure substance will conduct electricity only when melted?

    a) MgI_2 b) Xe   c) CBr_4 d) Pb

5. At the **triple point** of a substance, all three states are in dynamic equilibrium. The triple point of water is at 4.6 torr and 0.01°C. Which has the highest kinetic energy at the triple point?

    a) $H_2O(s)$ b) $H_2O(l)$ c) $H_2O(g)$
    d) they all have the same kinetic energy

6. Which substance in question 5 has the highest potential energy at the triple point?

7. Which is not increased by increasing temperature?

    a) volume of a gas   b) volatility of a gas
    c) pressure of a gas at constant volume
    d) all are increased

8. Which is not increased by increasing atmospheric pressure?

    a) boiling point
    b) solubility of a component gas in water
    c) volume of a gas at constant temperature
    d) all are increased

9. A gas has a volume of 25.0 liters at 25°C. At what temperature will it have a volume of 75.0 liters if the pressure is held constant?

    a) 75°C   b) 621 °C   c) 8.3 °C   d) -174 °C

10. A gas has a volume of 32.0 liters at 450 torr. What is the volume of the gas at 900 torr if the temperature is held constant?

    a) 16.0 L   b) 64.0 L   c) 0.0625 L   d) 12,700 L

11. A gas has a volume of 15.5 liters at 50°C and 110 torr. What is its pressure at 10°C in a 28.0 liter container?

    a) 15.2 torr   b) 174 torr   c) 53.4 torr
    d) 60.9 torr

12. A mixture of He and Xe has a pressure of 2.75 atmospheres. The pressure of the Xe is 1.60 atmospheres. What is the pressure due to the He?

    a) 874 torr   b) 0.00151 torr   c) 3306 torr
    d) 442 torr

ANSWERS TO SELF-TEST

Fill in the Blanks

1. metallic   2. increases
3. equilibrium vapor pressure or rate of evaporation; cohesive
4. melting
5. equilibrium vapor pressure; atmospheric pressure
6. less (has less mass)   7. increases   8. molecular
9. cohesive; disruptive   10. condensation

11. dynamic equilibrium   12. increases
13. higher; lower   14. amorphous

True or False

1. False--only depends on substance and temperature
2. True--ionic compound   3. True
4. False--molecular crystal since covalent compound
5. True--a larger pressure causes a smaller volume
6. True
7. True--same kinetic energy so He with a smaller mass has a higher velocity
8. True--check with Charles' law
9. True--gases are always more compressible than liquids
10. True-- 760 torr = 1 atm

Short Answer

1. 0.434 atm   (330/760)   2.   243 K   [273 + (-30)]
3. metallic   4. molecular
5. 0.90 J (1/2 x 0.20 x 3.0^2)
6. 14.7 L (12.0 x 353/288)   7.   197 torr (0.200 x 985)
8. 3.50 atm (14.0 x 3.00/12.0)
9. 1.37 atm (1.00 x 373/273)
10. 1.20 atm (790 torr is 790/760 = 1.04 atm)
11. a) 38°C   b) 80°C
12. a) 4.58 torr   b) 92.51 torr

Multiple Choice

1. d (1/2 x 7.0 x 10.0^2)   2. b (Ca is the only metal)
3. c (covalent compound; the only molecular crystal)
4. a (ionic compound)
5. d (same temperature; same kinetic energy)
6. c (have the same kinetic energy, but had to supply heat to achieve the gaseous state)
7. d
8. c (volume decreases as pressure increases)
9. b (298 x 75/25)
10. a (32.0 x 450/900)
11. c (110 x 15.5/28.0 x 283/323)
12. a [(2.75 - 1.60)x 760)

8

SOLUTIONS OF ACIDS, BASES, AND SALTS

CHAPTER OVERVIEW

This chapter discusses solutions and also mentions colloidal dispersions and suspensions. Solutions of acids, bases and salts as well as their properties and uses are stressed. The concentration units of percent by weight-volume and molarity are used. The extent of ionization is described as being the difference between strong and weak electrolytes. The pH logarithmic scale is used as a measure of a solution's acidity or basicity.

TOPIC SUMMARIES AND EXAMPLES

Introduction

A solution is a homogeneous mixture, composed of the **solvent**, which is the dissolving medium, and the **solute**, which is the dissolved substance. When both substances have the same state, the one present in greater amount is the solvent.

Example 8.1

Which is the solvent and which is the solute in the following situations?

   a) gaseous NH_3 added to water makes a solution which is liquid in appearence
   b) solid KI is dissolved in water to make a solution which is liquid in appearence
   c) 20 milliliters of liquid alcohol is added to 50 milliliters of liquid CCl_4 to make a solution which is liquid in appearance

Solution

    a) NH_3-solute; water-solvent
    b)   KI-solute; water-solvent
    c)   alcohol-solute; CCl_4-solvent

True Solutions, Colloidal Dispersions and Suspensions

True solutions have tiny solute particles which cannot be seen as being distinct from the solvent. The solute will not settle out and the solution can pass through either a filter or a semipermeable membrane. In a **colloidal dispersion** or **colloid**, the dispersed species has a larger particle size. It will cause the **Tyndall effect** in which a beam of light shined through the colloid gives a cloudy appearence. In addition, a colloidal dispersion will pass through a filter, but not through a semipermeable membrane. **Suspensions** are mixtures from which particles settle out upon standing. Use of a filter separates the suspended material from the other medium. Table 8-2 in the text summarizes these properties.

Concentrations of Solutions

Solute may be added to a solvent until no more will dissolve. At that point, the solution is said to be **saturated** with solute. The amount of solute present in any solution, saturated or unsaturated, can be expressed in many ways. The text mentions percent by weight-volume and molarity.

Percent by Weight-Volume

In this concentration unit, the solute's weight is commonly expressed in grams, while the total solution volume is reported in milliliters. The formula for **percent by weight-volume (%w/v)** is:

$$\text{percent by weight-volume} = \frac{\text{grams of solute}}{\text{milliliters of solution}} \times 100$$

Example 8.2

What is the percent by weight-volume of a solution made by adding water to 20.0 grams of glucose to make a solution with a volume of 400 mL?

Solution

$$? \text{ \% w/v} = \frac{20.0 \text{ g glucose}}{400 \text{ mL solution}} \times 100 = \mathbf{5.00\%}$$

Example 8.3

How many grams of NaCl must be mixed with water to prepare 500 milliliters of a 0.90 percent by weight-volume solution of NaCl?

Solution

The %w/v formula can be algebraically rearranged to give:

$$g\ solute = \%w/v\ \times\ \frac{mL\ solution}{100}\ ;\ \ therefore:$$

$$?\ g\ NaCl = 0.90\ \%\ \times\ \frac{500\ mL\ solution}{100} = 4.5\ g\ NaCl$$

Molarity

Molarity (M) is the most useful concentration unit to a scientist because it relates moles to easily measurable solution volumes. Review the concept of moles and formula weights from Chapter 2 if necessary. Molarity is the number of moles of solute in one liter of solution or:

$$Molarity = M = \frac{moles\ of\ solute}{1\ liter\ of\ solution} = \frac{mol}{L}$$

Molarity can be computed by dividing the number of moles of solute by the total solution volume in liters.

Example 8.4

What is the molarity of NaCl of the solution prepared in the previous example?

Solution

The solution was made from 4.5 grams of NaCl in 400 milliliters of solution. The formula weight of NaCl is 23 + 35 = 58 g/mol.

$$?\ mol\ NaCl = 4.5\ g\ NaCl\ \times\ \frac{1\ mol\ NaCl}{58\ g\ NaCl} = 0.078\ mol\ NaCl$$

$$?\ L\ of\ solution = 400\ mL\ solution\ \times\ \frac{1\ L}{1000\ mL} = 0.400\ L$$

The molarity can now be found:

$$? \ M = \frac{\text{moles solute}}{\text{liters of solution}} = \frac{0.078 \ \text{mol NaCl}}{0.400 \ \text{L solution}} = \textbf{0.20 M NaCl}$$

Example 8.5

How many grams of KBr are needed to prepare 250 milliliters of 0.200 M KBr solution?

Solution

The formula weight of KBr is 39 + 80 = 119 g/mol. The relationship that M = mol/L can be rearranged algebraically to give mol = M x L, then:

$$? \ \text{mol KBr} = 250 \ \text{mL} \ x \ \frac{1L}{1000 \ \text{mL}} \ x \ \frac{0.200 \ \text{mol}}{1 \ \text{L}} = 0.0500 \ \text{mol KBr}$$

$$? \ \text{g KBr} = 0.0500 \ \text{mol KBr} \ x \ \frac{119 \ \text{g KBr}}{1 \ \text{mol KBr}} = \textbf{5.95 g KBr}$$

Classes of Inorganic Compounds: Acids, Bases, Salts and Oxides

Inorganic compounds fall into four main categories. Some simple examples and distinguishing characteristics of each follow. **Acids** release hydrogen ions (H^+) in solution. **Bases** release hydroxide ions (OH^-) in solution. A **salt** is formed from the cation of a base and the anion of an acid and is the neutralization product of an acid and a base. **Oxides** are compounds in which an element is combined with only oxygen.

Properties of Acids

Acids impart a sour taste to foods and beverages. Acids can change the color of certain dyes called indicators and they can react with active metals to produce hydrogen gas in a reaction such as:

$$2HBr(aq) + Zn(s) \ \text{-->} \ ZnBr_2(aq) + H_2(g)$$

(Review single replacement reactions in Chapter 5.)   Acids can neutralize bases and they can dissolve in water to produce H^+ ions as in the following ionization process:

$$HCl(g) \ \xrightarrow{\text{water}} \ H^+ \ (aq) + Cl^- \ (aq)$$

The H^+ species becomes attracted to the partially negative oxygen end of the polar water molecule. The association is so strong that it is stated that the H^+ actually bonds to the water molecule and creates a **hydronium ion** (H_3O^+). The ionization above would then be rewritten as:

$$HCl(g) + H_2O(l) \longrightarrow H_3O^+(aq) + Cl^-(aq)$$

A Look at the Discovery and Uses of Acids

Vinegar's history is described and it is mentioned as being composed of one of the organic acids, acetic acid, $HC_2H_3O_2$. Later, **mineral acids** such as sulfuric acid were formulated. The book lists some commonly employed acids. You should study the characteristics and uses of: hydrochloric acid (HCl, also called muriatic acid), nitric acid (HNO_3), sulfuric acid (H_2SO_4), phosphoric acid (H_3PO_4), acetic acid ($HC_2H_3O_2$) and boric acid (H_3BO_3).

Properties of Bases

Bases impart bitter tastes to food and feel slippery. They change the color of indicator dyes to a different color from that caused by acids. Bases produce OH^- in solution. Most common bases are metal hydroxides which are ionic compounds. They dissolve in water because of the polarity of the water molecule. A sample ionization is:

$$Ba(OH)_2(s) \xrightarrow{\text{water}} Ba^{2+}(aq) + 2OH^-(aq)$$

Some Bases and Their Uses

You should be familiar with the uses and characteristics of the bases listed in the text: sodium hydroxide (NaOH, also called lye), calcium hydroxide ($Ca(OH)_2$, also called lime water), magnesium hydroxide ($Mg(OH)_2$, also called milk of magnesia) and ammonia (NH_3). The latter does not have a hydroxide ion in its formula, but instead produces it by the following equilibrium reaction:

$$NH_3(g) + H_2O(l) \Longleftrightarrow NH_4^+(aq) + OH^-(aq)$$

Properties of Salts

The double replacement neutralization reaction of an acid and a base produces a **salt** and water. In a complete neutralization reaction, all of the acidic hydrogens and all of the hydroxide ions must be converted to water. The salt formed should be electrically neutral. You should review oxidation numbers and charges of polyatomic ions in Chapter 4 to be certain you are writing the proper formula of the salt.

Example 8.6

Write the neutralization reactions for the following acid/base pairs.

    a) HBr and KOH   b) H_2SO_4 and LiOH

Solution

    a) HBr has one H^+ and KOH has one OH^-. The charge of K in compounds is +1, while that of Br is -1.

        $HBr(aq) + KOH(aq) \longrightarrow KBr(aq) + H_2O(l)$

    b) H_2SO_4 has two H^+ and LiOH has one OH^-. The charge of a sulfate ion is -2, while that of a Li ion is +1.

        $H_2SO_4(aq) + 2LiOH(aq) \longrightarrow Li_2SO_4(aq) + 2H_2O(l)$

In the previous example, the salts are the KBr and Li_2SO_4 which are formed.

Some Salts and Their Uses

The book discusses calcium phosphate, iron, sodium and potassium salts. A summary of some salts with their medical uses appears in Table 8-5.

Ionization and the Concept of Electrolytes

Solutions of sufficient concentrations of **electrolytes** in water conduct electricity, while those of **nonelectrolytes** do not conduct electricity at any concentration. Electrolytes are substances that produce ions which can carry current when they are dissolved. The breakup into ions is called **ionization**. Acids, bases and salts are all considered to be electrolytes, because they will produce ions when dissolved in water. Sample acid and base ionization equations appear earlier in this discussion. A salt will ionize to form its characteristic ions. For example, the Li_2SO_4 shown in Example 8.5b ionizes as follows:

    $Li_2SO_4(s) + H_2O(l) \longrightarrow 2\ Li^+(aq) + SO_4{}^{2-}(aq)$

Strong and Weak Electrolytes

Strong electrolytes ionize essentially completely in solution. Consult Table 8-6 in the text for some examples. Strong acids and bases are strong electrolytes. This means that if sufficient HCl, a strong acid, is placed in water to make a 2 M solution, it will totally ionize. The solution will contain 2 M H_3O^+ (H^+) ions and 2 M Cl^- ions and <u>no</u> HCl molecules. **Weak** electrolytes ionize to only a small extent in solution. In a solution containing a weak electrolyte, a dynamic equilibrium becomes established between the unionized molecule and its ions. In other words, at some point, the rate of ionization becomes equal to the rate of recombination of the ions. Equilibria are shown with a double arrow to show that both processes are occurring. Weak acids and bases are weak electrolytes. For example, HF is a weak electrolyte:

$$HF \ (aq) + H_2O \ (l) \ <===> \ H_3O^+ \ (aq) + F^- \ (aq)$$

In a 2 M solution of HF, after equilibrium is established, the HF has a molarity of about 1.96 M and the H_3O^+ and F^- each have molarities of about 0.04 M. A strong electrolyte always provides a higher concentration of ions than does a weak electrolyte with the same initial molarity.

The Ionization of Water

Water has a tendency to ionize to a certain extent, but the degree of ionization is so small that pure water cannot conduct electricity. The ionization can be shown as:

$$H_2O(l) + H_2O(l) \ <===> \ H_3O^+(aq) + OH^-(aq)$$

or

$$H_2O(l) \ <===> \ H^+(aq) + OH^-(aq)$$

In pure water at $25^{\circ}C$, the molarity of water is 55.6 M while the concentration of H^+ and OH^- is each only 0.0000001 M. Table 8-9 in the text shows the relationship of exponential notation to decimal notation. For example, the above molarity of OH^- or H^+ is 10^{-7}. Often square brackets are used to indicate the molarity of the substance of interest. $[H_3O^+]$ means the concentration of the hydronium ion in mol/liter.

Adding an acid to water causes the $[H_3O^+]$ to increase and the $[OH^-]$ to decrease, while the addition of a base has the opposite effect.

The pH Scale

The **pH scale** is designed as a measure of the acidity or basicity of a solution. It is a logarithmic scale:

$$pH = - \log [H_3O^+] \ \text{ or } \ pH = -\log [H^+]$$

In pure water, the pH is $-\log (10^{-7}) = -(-7) = 7$. This is considered to be a **neutral** pH. Note that when written in exponential form, the log of a number is simply the exponent of ten. In an acidic solution, the hydronium ion concentration is greater than 10^{-7} and the pH is less than 7. In a basic solution, the $[H_3O^+]$ is less than 10^{-7} and the pH is greater than 7. Since the pH scale is a logarithmic one, each gradation on the scale is a factor of ten. Thus, a solution with a pH of 4 has ten times the $[H_3O^+]$ as does a solution with a pH of 5 and ten times ten or one hundred times the $[H_3O^+]$ present in a solution with a pH of 6. Table 8-8 in the text gives some pH values for common substances. The text also gives guidelines for pH values of strongly acidic to strongly basic solutions.

The Relationship Between Hydrogen Ions and Hydroxide Ions

In water solutions, the product of the hydronium ion concentration and the hydroxide ion concentration gives a constant value called the **ionization constant** of water. At $25^\circ C$, their product is:

$$[H_3O^+][OH^-] = [H^+][OH^-] = 10^{-14}$$

If either the $[H_3O^+]$ or the $[OH^-]$ is known, the other can be calculated from this relationship.

Example 8.7

Find the $[H_3O^+$, the $[OH^-]$ and the pH of the following solutions. Also, tell if each is acidic or basic.

   a) 10^{-4} M HNO_3 b)   1 M KOH   c) 10^{-2} M HCl
   d)   0.00001 M NaOH

Solution

   a) HNO_3 is a strong acid, so the $[H_3O^+] = 10^{-4}$ M

      The pH = $-\log(10^{-4}) = -(-4) = 4$ (moderately acidic)

      The $[OH^-] = 10^{-14}/10^{-4} = 10^{-10}$ M

   b) KOH is a strong base, so the $[OH^-] = 1$ M

      The $[H_3O^+] = 10^{-14}/1 = 10^{-14}$ M

      The pH = $-\log(10^{-14} = -(-14) = 14$ (very basic)

   c) HCl is a strong acid, so $[H_3O^+] = 10^{-2}$ M

      The pH = $-\log(10^{-2}) = -(-2) = 2$ (very acidic)

      The $[OH^-] = 10^{-14}/10^{-2} = 10^{-12}$ M

d) NaOH is a strong base, so $[OH^-]$ = 0.00001 M = 10^{-5} M

The $[H_3O^+]$ = $10^{-14}/10^{-5}$ = 10^{-9} M

The pH = -log (10^{-9}) = -(-9) = 9 (slightly basic)

SELF-TEST

Fill in the Blanks

1. The reaction of an acid with a base is a/an _____ reaction and it produces a/an _____ and _____.

2. A/an _____ electrolyte only partially ionizes in solution, while a/an _____ electrolyte totally ionizes.

3. _____ are mixtures from which particles settle out as the mixture stands.

4. In a mixture of instant coffee and water, the _____ is the solvent and the _____ is the solute.

5. A/an _____ solution can have no more of that solute dissolved in it.

6. A base produces _____ ions when dissolved in water.

7. The $[OH^-]$ in a neutral solution is _____.

8. The pH of a solution in which the $[H_3O^+]$ is 10^{-11} M is _____.

9. _____ exhibit the Tyndall effect.

10. A solution with pH = 2 has _____ times the concentration of $[H_3O^+]$ as a solution with pH = 6.

True or False

1. True solutions will not pass through a semipermeable membrane.

2. In a mixture of acetic acid (a liquid) and water, the solvent is the substance present in greater amount.

3. The pH of a 10^{-6} M solution of HCl is 6.

4. When 2.0 moles of KOH are dissolved in 100 mL of water, the molarity is 0.0020 M.

5. A 1 M solution of a strong acid has a lower pH than a 1 M solution of a weak acid.

6. Bases taste sour.

7. A solution with $[OH^-] = 10^{-4}$ is acidic.

8. HCl is "stomach acid".

9. The salt formed by neutralization of $Ca(OH)_2$ with HBr is CaBr.

10. The $[OH^-]$ in a solution with pH = 8 is 10^{-6} M.

Short Answer

1. Write the ionization reactions for the following substances.

   a) HI (strong acid)   b) HCN (weak acid)
   c) LiOH (strong base) d) $HClO_4$ (strong acid)

2. Write the complete neutralization reactions for the following acid/base combinations.

   a) HNO_3 and $Ca(OH)_2$ b) H_2S and KOH
   c) H_3PO_4 and $Ba(OH)_2$ d) $Sr(OH)_2$ and H_2SO_4

3. What is the percent by weight-volume of NaOH in a solution made from 8.0 g NaOH dissolved to make 250 mL of total solution?

4. What is the molarity of NaOH in the solution in question 3?

5. How many grams of NaCl must be used to make 750 milliliters of 2.5 percent by weight-volume NaCl solution?

6. How many milliliters of 0.20 M HCl solution can be made from 1.8 grams of HCl?

7. What is the pH of a solution in which the hydronium ion concentration is 10^{-2} M?

8. What is the $[H_3O^+]$ in a 0.010 M solution of KOH?

9. A hair remover has a pH of 11.3. Is this solution acidic or basic?

Multiple Choice

1. Which is not a characteristic of true solutions?

   a) homogeneous   b) passes through a filter
   c) settles out
   d) passes through a semipermeable membrane

2. The salt formed when H_3PO_4 is neutralized by LiOH is:

   a) $LiPO_4$ b) $LiPO_3$ c) Li_2P d) Li_3PO_4

3. The weak base used in smelling salts is:

   a) NaOH   b) $Mg(OH)_2$ c) $Ca(OH)_2$ d) NH_3

4. "Battery acid" is:

   a) HCl   b) HNO_3 c) H_2SO_4 d) H_3BO_3

5. An acid used for flavoring drinks is:

   a) HNO_3 b) H_3PO_4 c) H_3BO_3 d) HCl

6. Which of the following is a salt?

   a) KNO_3 b) CuO   c) HNO_2 d) $Al(OH)_3$

7. What is the percent by weight-volume of HNO_3 in 500 milliliters of solution which contains 7.0 grams of HNO_3?

   a) 0.22 %   b) 71 %   c) 1.4 %   d) 35 %

8. What is the molarity of the solution in question 7?

   a) 22 M   b) 0.22 M   c) 0.014 M   d) 18 M

9. Which solution is the most acidic?

   a) one with pH = 3   b) 0.0001 M HI (strong acid)
   c) one with $[OH^-] = 10^{-12}$ M
   d) one with $[H_3O^+] = 10^{-5}$ M

10. Which pH represents a solution with 1000 times more $[OH^-]$ than a solution with pH = 5?

    a) pH = 1   b) pH = 3   c) pH = 7   d) pH = 8

93

ANSWERS TO SELF-TEST

Fill in the Blanks

1. neutralization; salt; water   2. weak; strong
3. suspensions   4. water; instant coffee
5. saturated   6. hydroxide or $[OH^-]$ 7. 10^{-7}
8. 11   9. colloidal suspensions   10. 10,000

True or False

1. False   2. True   3. True   4. False--2.0/0.100
5. True--strong acid makes more H_3O^+ per mole
6. False   7. False--the pH is 8   8. True
9. False--$CaBr_2$ 10. True

Short Answer

1. a) $HI(aq) + H_2O(l) \longrightarrow H_3O^+(aq) + I^-(aq)$

   b) $HCN(aq) + H_2O(l) \Longleftrightarrow H_3O^+(aq) + CN^-(aq)$

   c) $LiOH(s) \longrightarrow Li^+(aq) + OH^-(aq)$

   d) $HClO_4(aq) + H_2O(l) \longrightarrow H_3O^+(aq) + ClO_4^-(aq)$

2. a) $Ca(OH)_2 + 2HNO_3 \longrightarrow Ca(NO_3)_2 + 2H_2O$

   b) $H_2S + 2\ KOH \longrightarrow K_2S + 2H_2O$

   c) $2H_3PO_4 + 3Ba(OH)_2 \longrightarrow Ba_3(PO_4)_2 + 6H_2O$

   d) $Sr(OH)_2 + H_2SO_4 \longrightarrow SrSO_4 + 2H_2O$

3. 3.2 %   (8.0/250 x 100)
4. 0.80 M   [(8.0/40)/0.250; FW of NaOH = 40]
5. 19 g NaCl   (750 x 2.5/100)
6. 250 mL   [{(1.8/36)/0.20} x 1000; FW of HCl = 36]
7. 2   ($-\log 10^{-2}$)   8. 10^{-12} ($10^{-14}/10^{-2}$)
9. strongly basic

Multiple Choice

1. c   2. d   3. d   4. c   5. b   6. a
7. c   [(7/500) x 100]   8. b   [(7.0/63)/0.500]
9. c   10. d

9

NUCLEAR CHEMISTRY

CHAPTER OVERVIEW

This chapter discusses the discovery of radioactivity and the three major types of radiation. By emitting alpha or beta particles, one isotope can change into another. The concept of half-life is introduced and then used in a discussion of radioactive dating of objects. Fission and fusion are described. The adverse effects, the beneficial uses and the means of monitoring radiation are presented.

TOPIC SUMMARIES AND EXAMPLES

Introduction

This section describes some of the benefits of the use of radiation. It also mentions that radiation can be deadly and presents a true scenario of the effects and aftermath of improperly handled radioactive material.

History

The discovery of X-rays and aspects of natural radio-activity are presented along with the names of their discoverers. Natural radiation is found to be composed of three major types of radiation.

Alpha Particles, Beta Particles and Gamma Rays

Gamma rays (γ) are highly energetic and can penetrate thick walls. They have neither charge nor mass. **Alpha particles** (α) have the charge and mass of a helium nucleus, ^4_2He. They will not penetrate skin and are stopped by a sheet of paper. **Beta particles** (β) have the charge and mass of an electron. They can be stopped by a thin sheet of metal.

Beta particles arise from the breakdown of a neutron in the nucleus:

$$^1_0n \longrightarrow {}^1_1p + \beta \qquad \text{where } \beta = {}^0_{-1}e$$

Review isotopic designations in Chapter 2. Remember that the superscript is the mass number--the sum of the neutrons and protons. The subscript is the atomic number, which is the number of protons. The neutron is given an atomic number of zero since it has neutral charge. The electron is given an atomic number of -1 since it has the opposite charge from that of a proton. Note that the sum of the mass numbers and the atomic numbers is the same on either side of the nuclear equation. In nuclear equations, mass and charge are conserved rather than specific elements. **Transmutation**, which is the change from one element to another, is a major characteristic of nuclear reactions.

Natural Radioactivity

When an element releases one of the types of radiation, it is said to undergo **radioactive decay** or **nuclear decay**. When an alpha or a beta particle is emitted, the isotope undergoes **nuclear transformation** creating a different element.

Example 9.1

What are the isotopes formed when:

a) $^{189}_{75}Re$ releases a beta particle?

b) $^{230}_{90}Th$ releases an alpha particle?

c) $^{146}_{62}Sm$ releases an alpha particle?

d) $^{20}_{10}Ne$ releases a beta particle?

Solution

Remember that the atomic numbers and the mass numbers must be conserved. The symbol of the element formed must agree with the atomic number.

a) $^{189}_{75}Re \longrightarrow {}^0_{-1}e + {}^{189}_{76}Os$

b) $^{230}_{90}Th \longrightarrow \,^{4}_{2}He \,+\, ^{226}_{88}Ra$

c) $^{146}_{62}Sm \longrightarrow \,^{4}_{2}He \,+\, ^{142}_{60}Nd$

d) $^{24}_{10}Ne \longrightarrow \,^{0}_{-1}e \,+\, ^{24}_{11}Na$

Half-Life

The **half-life** of a radioactive isotope is the length of time required for one half of the atoms or one half of the mass of an isotope to decay. Each isotope has a specific half-life which can be found in tabulated data.

Example 9.2

The half-life of samarium-143 is 1.0 minute. If we have 20.0 grams of samarium-143 now, how much would be left in 5.0 minutes?

Solution

Half of the mass decays each 1.0 minute, therefore half of the previous amount is left at the end of each minute.

Time Elapsed	Amount Remaining
Now	20.0 g
1.0 min	10.0 g
2.0 min	5.00 g
3.0 min	2.50 g
4.0 min	1.25 g
5.0 min	**0.625 g**

Example 9.3

A sample of silver-103 is studied for 4.0 hours. It is found that its radioactivity has decreased to 1/16 of its original value at the end of that time. What is the half-life of this isotope?

Solution

If the sample has 1/16 of the radioactivity, it also must have 1/16 of the mass. We can set up a table as in the previous example.

Time Elapsed	Amount Remaining
Now	mass
1 half-life	1/2 mass

2 half-lives	1/4 mass
3 half-lives	1/8 mass
4 half-lives	1/16 mass

Four half-lives are needed for the mass and level of radioactivity to decrease to 1/16 of its former value. The length of its half-life must be 4.0 hours/4 = **1.0 hour**.

Using Radioactive Dating

Radioactive techniques can be used to estimate the age of various substances. The two most often consulted methods are to use the **decay series** of uranium-238 which eventually creates lead-206. The relative amounts of the lead and the uranium can be monitored. Half-lives are known. The method is an estimate since it is assumed that the sample originally contained no lead. Carbon-14 is used to determine the age of objects that were once living. Living objects incorporate carbon-14 into their tissues. After their death the carbon-14 decays, but the amount of carbon-12 stays constant. By examining the ratio, the age of the object can be approximated. This method is not exact because the amount of carbon-14 present in the atmosphere has been shown to vary.

Example 9.4

The half-life of carbon-14 is 5730 years. A cotton cloth sample is found to have 1/4 the amount of ^{14}C relative to ^{12}C as is found in modern cotton. What is the age of the cloth?

Solution

Consulting Example 9.3 shows that when the amount of material is 1/4 the original amount, two half-lives have passed. The cloth is 2(5730) or **11,460 years old**.

Humanly Produced Radiation

Rutherford first produced a man-made isotope by bombarding particles with alpha rays. The Curies then produced the first artificial radioactive isotope in the same manner. The elements on the periodic chart with mass numbers higher than that of uranium have been produced by similar techniques. All of these man-made elements have proven to be radioactive.

Example 9.5

What are the isotopes produced as the result of the following human experiments?

a) $^{27}_{13}Al$ + $^{4}_{2}He$ --> _____ + $^{1}_{0}n$

b) $^{253}_{99}Es$ + $^{4}_{2}He$ --> _____ + $^{1}_{0}n$

Solution

Remember that mass numbers and atomic numbers must be conserved. The symbol must agree with the atomic number. The proper products are:

a) $^{30}_{15}P$ b) $^{256}_{101}Md$

Nuclear Fission

Occasionally when bombarding isotopes with particles, they break up into less massive atoms instead of creating heavier nuclei. This breakup is called **fission**. Fission reactions are ordinarily initiated by neutron bombardment and produce additional neutrons and a tremendous amount of energy as a product. The production of additional neutrons leads to a **branching chain reaction** which is the basis of the atomic bomb. An example of fission is:

$^{235}_{92}U$ + $^{1}_{0}n$ --> $^{94}_{38}Sr$ + $^{139}_{54}Xe$ + 3 $^{1}_{0}n$

The Atom and Electric Power

The energy released in fission reactions, primarily with uranium-235, has been harnessed and used for electrical generation in nuclear power plants. The book mentions several problems which have occurred when controls have not been properly maintained and radiation has been released into the environment. Three other problems with the use of fission as a fuel are discussed. Along with energy, a huge mass of byproducts is produced. Many of the byproducts are themselves radioactive with long half-lives. These present a massive disposal problem. Additionally, world reserves of fissionable uranium are being rapidly depleted, so plants may have to be closed. Finally, considerable heat pollution occurs when the large volume of water needed to cool the power plant is discharged back into bodies of water.

The Breeder Reactor

A **breeder reactor** causes the conversion of abundant non-fissionable uranium-238 to fissionable plutonium-239 and then uses the plutonium-239 as fuel. This eliminates the problem of running out of fissionable material but introduces additional problems of security. Plutonium-239 is highly toxic and it is relatively simple to convert it into atomic bombs. These factors and the problems of heat control have prevented wide-spread use of breeder reactors.

Nuclear Fusion

Nuclear fusion is the opposite of fission. In this process, which powers the sun, lighter nuclei combine to form heavier atoms with the release of energy. Controlled fusion has not been sustained on the earth, but fusion has been employed in the hydrogen bomb. Advantages of fusion over fission would be that the hydrogen isotope **deuterium** (^{2}H) which is required is readily available. Additionally, the byproducts would not be radioactive. We would only need to contend with thermal pollution.

Radiation and Its Medical Uses

Radiation can cause cellular damage as the particles or rays can actually break up and destroy cell tissue. Interactions of the radiation with other substances present such as water can create **free radicals** or hydrogen peroxide which can in turn do chemical damage to the cells. The cells can be prevented from undergoing normal functions, since reproduction is hindered and cancerous growths may be initiated. **Stage 1** effects are the actual changes in cell makeup as a result of the radiation, while **stage 2** effects are those effects which are the result of impaired cell function. An example of a stage 1 effect is the damaging of reproductive organs by radiation. A stage 2 effect is the miscarriages or sterility which result from the damage. Cells undergoing rapid growth such as in the reproductive organs, the bone marrow, children, fetuses, and cancers are most suscepticible to radiation damage because the radiation can alter the way in which these rapid reactions are occurring.

The Detection of Ionizing Radiation

The **Geiger-Muller counter**, the **scintillation counter**, **film badges** and **dosimeters** are described as being means by which radiation, either current (the first two methods) or cumulative (the second two methods), can be monitored.

Units Used for Expressing Dosages of Radiation

The **curie (Ci)** describes the activity of a radiation source and is equivalent to **3.7 x 10^{10} disintegrations per second (dps)**.

Example 9.6

A sample has 7.4×10^{10} disintegrations per second. How many curies is that?

Solution

$$? \text{ curies} = 7.4 \times 10^{10} \text{ dps} \times \frac{1 \text{ Ci}}{3.7 \times 10^{10} \text{ dps}} = 2.0 \text{ Ci}$$

A **roentgen (r)** measures ionizing ability and is the amount of radiation that produces 2×10^9 ion pairs in one cubic centimeter of air. The **rad (radiation absorbed dose)** is the amount of energy absorbed by living tissue, regardless of type of radiation. A **lethal dose** is the amount of radiation needed to kill an organism. Ordinarily lethal dosages are reported in LD_{50}^{30}, the amount of radiation which will kill 50 percent of the exposed organisms in 30 days. The **rem (roentgen equivalent man)** is equal to the radiation dosage in rads times a weighting factor. Table 9-5 in the text correlates the observed physiological effects to dosage received by a person.

Radioisotopes: Their Use in Medical Diagnosis and Therapy

Specific isotopes and their uses in the body are discussed. Iodine-131 is used for hyperthyroidism and monitoring of the thyroid. Chromium-151 is used to check for proper blood flow. Cobalt-59 and cobalt-60 are used to monitor vitamin B-12 absorption. Radiation is described as being used against cancerous tissues in three different methods. **Teletherapy** is the use of high energy beams targeted at the cancer. **Brachytherapy** is insertion of a radioactive sample into the body, in or near the tumor. **Radiopharmaceutical therapy** involves oral or intravenous insertion of the radioisotope. The isotope then follows normal metabolism or blood pathways to interact with the area requiring treatment.

SELF-TEST

Fill in the Blanks

1. _____ particles are the equivalent of a helium nucleus.

2. _____ have the most penetrating power of the natural radiations.

3. _____ effects involve changes in tissue structure.

4. A/an _____ counter determines the rate and energy of radioactivity.

5. The cells most affected by radiation are found in _____ tissues.

6. _____ is the breakdown of a heavier nucleus to lighter ones and _____ is the synthesis of more massive atoms from lighter ones.

7. The unit which measures ionizing ability is the _____.

8. When high energy beams of radiation are directed at a tumor, the treatment is called _____.

9. The fissionable isotope of uranium is _____ and the abundant isotope of that element is _____.

10. The natural radiation which has the charge and mass equivalent to that of an electron is _____.

11. A new element is not produced as a result of gamma radiation because a gamma ray has no _____ and no _____.

12. When a nuclear equation is balanced, _____ and _____ must be conserved.

True or False

1. 3.7×10^{10} Ci is a dps.

2. Fusion is used commercially in nuclear power plants.

3. A radiation dose of above 600 rems is usually fatal.

4. $^2_1H + ^2_1H \longrightarrow ^3_2He + ^1_0n$ is a fusion reaction.

5. The fact that radiation destroys bone marrow is a stage 1 effect.

6. Deuterium is used in breeder reactors.

7. Beta particles can be stopped by a sheet of paper.

8. A rem is a rad times a weighting factor.

9. When a beta particle is emitted, the product has a higher atomic number than the reactant.

10. Cobalt isotopes are used to check on thyroid activity.

Short Answer

1. Balance the following nuclear equations.

   a) $^{210}_{82}Pb \longrightarrow {}^{0}_{-1}e + $ _____

   b) $^{235}_{92}U + {}^{1}_{0}n \longrightarrow {}^{103}_{42}Mo + $ _____ $ + 2\ {}^{1}_{0}n$

   c) $^{9}_{4}Be + {}^{4}_{2}He \longrightarrow $ _____ $ + {}^{1}_{0}n$

   d) $^{243}_{95}Am \longrightarrow {}^{4}_{2}He + $ _____

   e) $^{246}_{96}Cm + {}^{12}_{6}C \longrightarrow $ _____ $ + 4\ {}^{1}_{0}n$

   f) $^{251}_{98}Cf \longrightarrow {}^{4}_{2}He + $ _____

   g) $^{238}_{92}U + {}^{2}_{1}H \longrightarrow $ _____ $ + {}^{1}_{1}H$

   h) $^{111}_{46}Pd \longrightarrow {}^{0}_{-1}e + $ _____

   i) $^{235}_{92}U + {}^{1}_{0}n \longrightarrow {}^{93}_{37}Rb + $ _____ $ + 2\ {}^{1}_{0}n$

   j) $^{40}_{19}K \longrightarrow {}^{0}_{-1}e + $ _____

2. Which of the reactions in question 1 are fission reactions?

3. The half-life of ^{184}Re is 50 days. If we start with 400 mg of that isotope now, about how much will be left in five months (about 150 days)?

4. Strontium-90 is one product of fission. It decays by beta emission with a half-life of 28 years. a) What is the product of its decay? b) How long does it take for 1000 pounds of this isotope to decay so that only 62.5 pounds remain?

5. ^{239}Am decays by alpha emission. a) What is the decay product? b) What is the half-life of this isotope in hours if a sample decays from 160 milligrams to 2.5 milligrams in 3 days?

6. How many disintegrations per second are in a sample which is a 0.30 curie emitter?

7. How old is a piece of wood if it has a $^{14}C/^{12}C$ ratio which is one eighth that of new wood?

8. List four types of tissue associated with particular susceptibility to radiation damage.

9. What are four problems associated with fission?

10. What would be two advantages of fusion over fission?

11. List six physiological results of exposure to radiation.

12. An organism has an LD^{30}_{50} of 250 rems. If 750 organisms are exposed to a 250 rem of radiation, how many of them should be dead in 30 days?

ANSWERS TO SELF-TEST

Fill in the Blanks

1. alpha   2. gamma   3. stage 1   4. scintillation
5. fast growing   6. fission; fusion   7. ˙roentgen
8. teletherapy   9. uranium-235; uranium-238
10. beta   11. mass; charge
12. mass number; atomic number

True or False

1. False--1 Ci = 3.7 x 10^{10} dps   2. False-fission
3. True   4. True   5. True   6. False--plutonium-239
7. False--alphas are stopped this way   8. True
9. True   10. False--iodine isotopes

Short Answer

1. a) $^{210}_{83}Bi$ b) $^{131}_{50}Sn$ c) $^{12}_{6}C$ d) $^{239}_{93}Np$

   e) $^{254}_{102}No$ f) $^{247}_{96}Cm$ g) $^{239}_{92}U$ h) $^{111}_{47}Ag$

i) $^{141}_{55}$Cs   j) $^{40}_{20}$Ca

2. b and i   3. 50 g   (400 x 1/2 x 1/2 x 1/2)

4. a) $^{90}_{39}$Y   b) 112 years (1000 x 1/2 x 1/2 x 1/2 x 1/2)

5. a) $^{235}_{93}$Np   b) 12 hours (160 x 1/2 six times)

6. 1.1 x 10^{10} (0.30 x 3.7 x 10^{10})
7. 17,190 years (three half-lives @ 5730 years apiece)
8. reproductive organs; bone marrow; fetuses and children; cancers; other fast-growing tissues
9. waste disposal; direct radiation leakage; thermal pollution; running out of fuel
10. fusion has readily available starting materials (reactants) and would have no waste disposal problem
11. nausea; loss of appetite; leukemia; cancers; mutations; diarrhea; fever; vomiting; miscarriages; death; etc.
12. 375 (half are dead in 30 days)

10

THE CHEMISTRY OF WATER AND WATER POLLUTION

CHAPTER OVERVIEW

This chapter studies the properties of water, its importance to life and the substances which pollute it. The ways in which water is polluted are traced. Treatments which are applied to make water potable and those which are used to remove contaminants from waste water are described.

TOPIC SUMMARIES AND EXAMPLES

Introduction

Potable water is described as being safe drinking water which must have a minimum of contaminants. This section mentions how DDT, a pesticide, and heavy metals have contaminated water systems. Phosphates from fertilizers are also a pollutant because they foster the growth of algae. Decay of the algae overgrowth further pollutes the water. In addition, algae absorb many substances which may be toxic or hazardous to us. The substances are then passed through the food chain to our detriment.

Physical Properties of Water

Pure water is a polar molecule that is colorless, tasteless and odorless. Water, unlike most other substances, has a liquid state that is more dense (1.0 g/mL) than its solid state (0.92 g/mL). The solid state floats upon the liquid state and helps to insulate it from cold air temperatures. This has the effect of keeping entire bodies of water from freezing and protects the aquatic life beneath the frozen surface. Water has this unique property because of the nature of its intermolecular attractions. Water exhibits **hydrogen bonding**, an intense attraction between the partially

negative oxygen on one water molecule and the partially
positive hydrogen on another water molecule. The molecules
can be positioned closer to each other in the randomly
oriented liquid state than they can when the water solidifies
into its characteristic crystals in the solid state.

Chemical Properties of Water

Water is a very stable molecule with a highly exothermic
heat of formation (-68.3 kcal/mol). It can be decomposed, if
an electrolyte is added to it, by electricity to yield
hydrogen and oxygen. Water reacts with nonmetal oxides to
form acids and with metal oxides to form bases.

Example 10.1

$N_2O_5 + H_2O \longrightarrow 2HNO_3$ (nitric acid)

$Li_2O + H_2O \longrightarrow 2LiOH$ (lithium hydroxide)

Some solids, called **hydrates**, exist with water molecules
incorporated into their crystal lattices. Heating these
hydrates causes the release of the incorporated water and
produces **anhydrous** salts.

Example 10.2

$$CuSO_4 \cdot 5H_2O(s) \xrightarrow{\text{heat}} CuSO_4(s) + 5H_2O(g)$$

Because of its polarity, water is found to be an
excellent solvent for a wide range of materials.
Unfortunately, its abilities as a solvent make it highly
susceptible to contamination, since unwanted as well as
desirable substances will dissolve in it.

The Water Cycle

This section traces the path of falling rain to the
ocean. In the ocean and along the way, water evaporates and
forms into clouds and the process can begin again. As water
flows to the ocean, it gets used by homes, agriculture and
industry for various purposes. Consider the number of times a
week you flush toilets, wash dishes or clothes and take
showers. These actions and such effects as rain runoff cause
water to become contaminated with sewage and other domestic
wastes, heavy metals and organic chemicals from industry, and
pesticides and fertilizers from agriculture.

A body of water has microorganisms which can cleanse the
water of a certain number of pollutants by oxidizing them to
water and carbon dioxide as they use them for energy. If the

water becomes overloaded by organic materials, such as food and human wastes, **aerobic** (oxygen needing) microorganisms such as algae can grow abundantly, nourished by these substances. Their growth consumes oxygen in the water, creating a **biological oxygen demand (BOD)**. Oxygen depletion causes the death of fish and other organisms needing oxygen for life. As these organisms die and begin to decay, the oxygen demand increases still more, accelerating the problem. **Anaerobic** (not oxygen needing) microorganisms then take over, producing as byproducts of their metabolism foul-smelling and toxic hydrogen sulfide gas, offensive ammonia vapors and potentially explosive methane (CH_4) gas.

Organic fertilizers, with nitrate and phosphate nutrients, hasten algae and other plant growth. As the plants die and decay, they fill in the body of water in a process called **eutrophication**. The death of the excess plant growth also hastens oxygen depletion by increasing the BOD. This in turn helps the takeover of the body of water by noxious anaerobic organisms.

Pesticides and industrial wastes can act as poisons and cancer-causing agents. Domestic wastes can provide disease agents, poisonous cleaning products and additional phosphates as well as anything else which is put down a drain.

Making Water Safe to Drink

Drinking water treatment was initiated to kill **pathogenic** organisms, which cause diseases such as cholera, typhus and hepatitis. **Chlorine** is the usual agent of choice to kill these organisms; however, it has been linked to cancer causing agents formed when it combines with organic chemicals. Because of this potential hazard, chlorine use must be carefully monitored. Potable water is often treated with coagulants to remove other impurities. Potable water should be free of: 1) turbidity (cloudiness); 2) color; 3) pathogens; 4) algae and other plants; 5) taste and odor; and, 6) phosphates, which promote growth of organisms in the water.

Home water may also be treated for "hardness" which is the presence of calcium, magnesium and iron(II) ions. Ca^{2+}, Mg^{2+} and Fe^{2+} are considered to be "hard" ions because they cause precipitation of a scum with soaps, inhibiting cleansing action. They also leave mineral deposits on such things as coffee pots. The iron ion is particularly noted for leaving "rust" rings on toilets and tubs. Water softeners usually work by exchanging "soft", very soluble sodium ions for the calcium, magnesium and iron(II) ions.

Wastewater Treatment: Saving Our Lakes, Rivers,and Streams

To protect bodies of water from domestic or industrial wastes, treatment of the effluent from these places is necessary before the water is discharged back into the environment.   Pollutants are categorized into **dissolved** materials and **undissolved** solids.   The undissolved solids are in turn subdivided into **settleable** and **suspended** solids. Typical domestic sewage has a BOD of about 200 milligrams of oxygen for each milliliter of water (200 mg/mL).   Although some water is still discharged without any treatment,   most water in the United States is subjected to some type of a purification process.

Primary Treatment

Primary treatment is a **physical** method of separation of removing settleable and some suspended solids from the water. A **bar screen** removes large particles and a **skimmer** removes suspended material from the surface.   The water is then allowed to stand in settling tanks.   After the settleable solids have settled, the water is drawn off the top and the solids are disposed.   Only about thirty percent of the pollutants are removed in this process, leaving a BOD of about 140 mg/mL.

Secondary Treatment

Secondary treatment is a **biological** method, which uses aerobic microorganisms to convert dissolved substances into stable inorganic substances, which can be settled out, or into carbon dioxide which bubbles harmlessly away.   One secondary treatment is the **activated sludge** process in which water is innoculated with a mass of aerobic bacteria and inorganic material, called sludge.   The system is aerated and the bacteria consume organic materials and create more sludge. The **trickling filter** system involves slowly passing water over rocks coated with aerobic bacteria.   Settling tanks are used in each case to let the byproducts precipitate from solution, so they can be removed from the cleansed water.   After this treatment, the BOD is characteristically about 20 to 40 mg/mL.

Advanced Treatment

Advanced treatment is a **chemical** method which removes specific contaminants for health or aesthetic reasons.   Some chemical is added to the water to remove a specific chemical pollutant.   For example, calcium is sometimes added to precipitate out phosphates to stop them from accelerating plant growth in the body of water into which the treated water will be discharged.   Advanced treatment tends to be expensive and is rarely done except for chlorination of the water. This usually is done as a final step to kill any pathogenic agents which may still be present.

SELF-TEST

Fill in the Blanks

1. Ice floats on liquid water because it has a _____ than does liquid water.

2. The process in which a body of water is filled in by plant material is called _____.

3. Metal oxides, when dissolved in water, make _____ solutions.

4. The strong attraction between different water molecules is called a/an _____ bond.

5. Organisms which do not need oxygen are said to be _____.

6. Secondary treatment is a/an _____ method of purifying water.

7. The "hard" water ions are _____, _____ and _____.

8. _____ is the method usually used to treat water to remove pathogens.

9. _____ are crystals which have water molecules incorporated into the crystal lattice. When they are heated the _____ salt and _____ are formed.

10. Undissolved solids in water are classified as _____ or _____.

True or False

1. Cl_2O_7 should make an acidic solution when it is dissolved in water.

2. Aerobic organisms produce noxious smells.

3. Water expands when it freezes.

4. Phosphates are considered to be a water pollutant because they are poisonous.

5. The activated sludge process is part of primary water treatment.

6. Primary water treatment lowers the BOD to about 140 mg/mL.

7. Water is easily polluted because it is a good solvent.

8. Pathogenic organisms are added to water to lower the BOD.

9. Because water is so easily contaminated, it is not considered to be a stable compound.

10. Use of chlorine has risks as well as benefits.

Short Answer

1. P_4O_{10} makes phosphoric acid when dissolved in water. Write the balanced equation for this process.

2. Write the equation for the breakdown of water into its elements.

3. Show the equation of the reaction that occurs when photographer's "hypo", $Na_2S_2O_3 \cdot 5H_2O$ is heated.

4. List six physical properties of pure liquid water.

5. List six common water pollutants.

6. List six common water conditions for which potable water is treated.

ANSWERS TO SELF-TEST

Fill in the Blanks

1. lower density   2. eutrophication   3. basic
4. hydrogen   5. anaerobic   6. biological
7. calcium (Ca^{2+}); magnesium (Mg^{2+}); iron(II) (Fe^{2+})
8. chlorination   9. hydrates; anhydrous salt; water
10. settleable; suspended

True or False

1. True--nonmetal oxide   2. False--anaerobic 3. True
4. False--they foster plant growth and hasten eutrophication
5. False--secondary   6. True   7. True

8. False--they are disease causing agents
9. False--very stable, but a good solvent
10. True--implicated in cancer causing byproducts

Short Answer

1. $P_4O_{10} + 6H_2O \longrightarrow 4H_3PO_4$

2. $2H_2O(l) \longrightarrow O_2(g) + 2H_2(g)$

3. $Na_2S_2O_3 \cdot 5H_2O(s) \longrightarrow Na_2S_2O_3(s) + 5H_2O(g)$

4. odorless, colorless, tasteless, the density of the liquid is about 1.0 g/mL, the boiling point is 100°C at 1.0 atm, the freezing (melting) point point is 0°C

5. pesticides, fertilizers, organic compounds, sewage, heavy metals, disease organisms, dead cows, disposable diapers, etc.

6. turbidity, color, odor and taste, pathogens, phosphates, algae, (plus, hopefully, everything else in the previous answer)

11

THE CHEMISTRY OF AIR POLLUTION

CHAPTER OVERVIEW

This chapter discusses the components of normal air and how concentrations of air pollutants are measured. Major air pollutants, their sources, and the hazards associated with them are presented. Information about pollution control devices is given, followed by a cost versus benefit discussion on the means of keeping air clean.

TOPIC SUMMARIES AND EXAMPLES

Introduction

This chapter opens with the definition of air pollution as being the presence of a contaminant in the atmosphere of sufficient concentration to injure life or interfere with its enjoyment. A scenario of the deaths due to smog in Donora, Pennsylvania is presented.

Measuring Pollution Concentration in the Air

Major components of gas mixtures are usually measured in percent by volume where:

$$\text{percent by volume} = \frac{\text{volume of substance of interest}}{\text{total volume of gas mixture}} \times 100$$

Example 11.1

To help prevent the bends, some divers now use a mixture of helium and oxygen because helium is less soluble in the blood than is nitrogen. What is the percent by volume of oxygen in a mixture of 8.0 liters of helium and 40.0 liters of oxygen?

Solution

The total volume is 8.0 L + 40.0 L = 48.0 L.

$$? \text{ \% by volume } O_2 = \frac{40.0 \text{ L } O_2}{48.0 \text{ L total}} \times 100 = \textbf{83\% } O_2$$

For minor components, concentrations are usually reported in **parts per million (ppm)**, where:

$$\text{ppm} = \frac{\text{volume of substance of interest}}{\text{total volume}} \times 1,000,000$$

Example 11.2

A sample of air is found to contain 8.0 milliliters of sulfur trioxide in 120 liters of air. What is the concentration of sulfur trioxide in parts per million?

Solution

The volumes must be in the same units, so milliliters are changed to liters.

$$? \text{ L } SO_3 = 8.0 \text{ mL} \times \frac{1L}{1000 \text{ mL}} = 0.0080 \text{ L } SO_3$$

Using the definition of parts per million,

$$? \text{ ppm } SO_3 = \frac{0.0080 \text{ L } SO_3}{120 \text{ L air}} \times 1,000,000 = \textbf{67 ppm } SO_3$$

Example 11.3

How many liters of sulfur dioxide are present in 500 liters of a sample which has an SO_2 concentration of 5.0 ppm?

Solution

Rearranging the definition of ppm algebraically yields:

$$? \text{ L } SO_2 = \frac{\text{ppm} \times \text{total volume}}{1,000,000} = \frac{(5.0 \text{ ppm})(500 \text{ L})}{1,000,000}$$

$$? \text{ L } SO_2 = \textbf{0.0025L or 2.5 mL}$$

Particulate matter in the air, such as soot, dust and liquid mists, is classified by the unit of micrograms per cubic meter ($\mu g/m^3$). A microgram is 0.000001 g (1×10^{-6} g).

Example 11.4

The concentration of soot in the air after a forest fire was found to be 0.00800 grams in 40.0 cubic meters of air. What is its concentration in $\mu g/m^3$?

Solution

The grams must be converted to micrograms.

$$?\mu g = 0.00800 \text{ g} \times \frac{1,000,000 \text{ g}}{1 \text{ g}} = 8,000 \ \mu g$$

$$? \ \mu g/m^3 = \frac{8,000 \ \mu g}{40.0 \ m^3} = 200 \ \mu g/m^3$$

The Composition of the Air We Breathe

Air is primarily composed of nitrogen and oxygen with other gases providing only about one percent by volume of the atmosphere. Table 11-1 in the text lists the components of ordinary dry air. Water vapor content varies greatly with location and temperature.

Major Air Pollutants: A Deep Deadly Breath

Table 11-2 in the text lists the pollutants that the Environmental Protection Agency (EPA) considers to be the major air pollutants. It also gives maximum acceptable levels of these pollutants. **Primary** pollutants are directly released into the air from some source. **Secondary** air pollutants are formed from chemical reaction of other substances in the air.

What We Know About Oxides of Sulfur

The major source of sulfur oxides is fuel combustion from the use of coal or fuel oil which usually contain sulfur impurities in varying amounts. Sulfur combines with oxygen to form **sulfur dioxide** (SO_2). The sulfur dioxide then can be oxidized further to **sulfur trioxide** (SO_3). These nonmetal oxides each react with water vapor to form **sulfurous acid** (H_2SO_3) and **sulfuric acid** (H_2SO_4). The gases themselves irritate and can damage the lungs and they also block light from the sun. The acids formed are a constituent of acid rain and can corrode marble and metals and kill plants.

What We Know About Carbon Monoxide

The major source of **carbon monoxide (CO)** is the internal combustion engine such as is found in automobiles. CO is formed when there is an incomplete supply of oxygen to totally burn fuels. With sufficient oxygen, non-toxic carbon dioxide is the end-product of combustion. Carbon monoxide is a pollutant which is colorless, odorless and tasteless, so it gives no irritating sign of its presence. It can move through the lungs directly into the blood stream where it preferentially binds with the hemoglobin of red blood cells with an affinity 200 times that of oxygen. Since the carbon monoxide takes the place of oxygen in the hemoglobin, cells, in essence, are smothered by being cut off from oxygen input. Drowsiness, followed by paralysis and death are results of sufficient concentrations of carbon monoxide.

What We Know About Photochemical Oxidants

Photochemical oxidants include **ozone** (O_3) and **peroxyacetyl nitrates (PANs)**. They are the ingredients of photochemical smog, found in warm, sunny climates. All photochemical oxidants are: 1) formed by **chemical** reaction; 2) form more rapidly in areas with **sunlight**; and, 3) act as **oxidizing agents**. Ozone is formed by a chain of events. At high temperatures, the nitrogen in the air can react with oxygen to form nitrogen monoxide. NO reacts with additional oxygen to make nitrogen dioxide, which is quite unstable and quickly degrades to nitrogen monoxide and an oxygen atom. The free oxygen atom then combines with ordinary atmospheric oxygen to form ozone. Not only is ozone a lung and skin irritant, but it quickly causes degradation of rubber and fabrics and kills vegetation. PANs are a combination of ozone, hydrocarbons and nitrogen oxides. They exhibit the similar physiological effects to those of ozone and, in addition, impart a brown color to the air.

What We Know About Hydrocarbons

The major source of **hydrocarbons** is the incomplete combustion of fuels in internal combustion engines. Hydrocarbons are sometimes released before they have fully reacted with oxygen and been converted to carbon dioxide and water vapor. Specific hydrocarbons can have adverse effects on vegetation; however, they are primarily considered to be a problem because of their subsequent conversion into PANs.

What We Know About Oxides of Nitrogen

The major source of nitrogen oxides is the combustion of fuels. Nitrogen is ordinarily a stable, relatively unreactive gas. At elevated temperatures, it can react with oxygen to form a variety of oxides, NO_x. In one example, **nitrogen monoxide (NO)** is formed by direct combustion. It can be further oxidized to **nitrogen dioxide (NO_2)**. The latter can react with water in the air to produce a mixture of **nitrous (HNO_2)** and **nitric (HNO_3)** acids. All nitrogen oxides are brown and reduce visibility. They destroy vegetation and cause corrosion of mtals. Some nitrogen oxides are highly toxic and can cause lung damage. Nitrogen oxides also react chemically to form PANs (the nitrate part).

What We Know About Particulate Matter

Particulate matter consists of solid dust and smoke and liquid mists and sprays. These come primarily from burning reactions and industrial and agricultural sprayings. Particulates irritate the lungs and block light from the earth. They may enhance the effects of other pollutants which they adsorb from the air. As the particles settle, the pollutants can react with the surface onto which they fall.

Hazardous Pollutants

Although the pollutants described previously do constitute a threat at high enough concentrations, the term hazardous pollutant, is reserved for species which accumulate in the body and cause problems as their level there increases.

What We Know About Asbestos

Asbestos is a fibrous mineral that has been widely used for insulation purposes, brake linings and protective clothing. The tiny particles can break off, lodging in the lungs and blocking air passages in a condition known as **asbestosis**. Lung cancer has been attributed to the presence of the fibers. Many buildings throughout the country are being treated to remove the old asbestos insulation and, thus, the threat of inhaling the substance.

What We Know About Beryllium

The major source of **beryllium (Be)** is the metal alloy and rocket fuel plants where it is used. Be can cling to workers and be transported to other sites. In the body, it can cause serious lung damage and a disease called **berylliosis**. The metal interferes with the biochemistry of the body, often causing death.

What We Know About Mercury

The major sources of **mercury (Hg)** are fuels and industry,

117

though broken thermometers and barometers may provide a source in many households. It has a high vapor pressure and can also be absorbed directly through the skin. Mercury has also been classified as a water pollutant and can move through the food chain, becoming concentrated to dangerous levels. Mercury particularly affects the brain and kidney function.

Reactions in the Air

Many reactions occur in air and some were mentioned earlier in the chapter. This section presents them in more depth and also discusses the "killer" smog that killed four thousand people in London in 1952.

Acid Rain

Acid rain is considered to be any precipitation which has a pH value of less than 5.6. The acids formed from the sulfur and nitrogen oxides are believed to be primarily responsible for this effect. Acid rains kill plant life, make the water impossible for animals to live in and corrode mineral and metal surfaces. Europe, eastern Canada, and the northeastern section of this country have been particularly hard hit by acid rain because of the industry in those areas. Relations with Canada are particularly sensitive because the Canadians attribute most of their problems with acid rain to American industry.

Problems with the Ozone Layer

Although surface ozone is considered to be a pollutant, the atmospheric ozone layer is of immense value. Ozone absorbs much of the ultraviolet radiation from the sun and dissipates it as heat. **Chlorofluorocarbons**, used as propellants, refrigerants and to make plastic foams have been found in the upper atmosphere. They apparently react with ultraviolet light to release free chlorine atoms. These very reactive atoms react with ozone to produce chlorine monoxide and diatomic oxygen. The chlorine monoxide is a reactive species which quickly degrades to form more chlorine atoms and the process continues. The ozone layer, especially over Antarctica, has been depleted significantly, so that there is now a "hole" in the ozone layer. The increased penetration of the ultraviolet light is expected to cause more sunburns, and, consequently, more skin cancer. The increase in ultraviolet light also does damage to plants.

The Dangers of Exposure to Radon Gas: An Indoor Air Pollutant

Various **radon (Rn)** isotopes are formed as the result of uranium decay series. The gaseous product can seep through the bases of houses where its radioactive daughter atoms can cling to airborn dust and become lodged in the lungs. The radioactivity then released has been implicated in lung cancer, with specific examples having been found in workers in the nuclear industry and in people living over uranium rich rock. Studies have found that sealing the base of houses and providing good ventilation, so that the radon is removed instead of building up in the house, is helpful.

How Can We Control Air Pollution

Emission of sulfur oxides can be decreased by the use of low sulfur fuels, and by using **scrubbers** which absorb gaseous pollutants. Particulate emissions can be decreased by the use of **electrostatic** precipitators and porous bags which can be penetrated by gases but not by the particles. **Catalytic mufflers** have been used to increase the efficiency of combustion in order to decrease formation of carbon monoxide and the discharge of hydrocarbons.

Air Pollution: A Social Commentary

About fifty percent of the air pollution in the country comes directly from the internal combustion engines of automobiles and other vehicles. This section discusses restricting vehicle use, the use of mass transit, and electric vehicles as possible solutions. It is stated that there are both economic and social consequences of the various proposed courses. For all future decisions, the book recommends considering both the costs and the benefits of environmental concerns.

SELF-TEST

Fill in the Blanks

1. _____ and _____ are the two major constituent gases found in clean air.

2. Two metals which are hazardous air pollutants are _____ and _____ .

3. The two major pollutants associated with internal combustion engines are _____ and _____ .

4. Two major pollutants responsible for acid rain are _____ and _____ .

5. The ozone layer is important because it absorbs _____ radiation.

6. _____ gas is a hazard because it is radioactive.

7. _____ and _____ are considered to be photochemical oxidants.

8. _____ have been associated with the destruction of the ozone layer.

9. If the pH of precipitation is below _____, it is considered to be acid rain.

10. _____ are one means of controlling gaseous emissions from smokestacks.

True or False

1. Carbon monoxide is associated with lung cancer.

2. Asbestos is another photochemical oxidant.

3. The automobile is not considered to be responsible for mercury pollution.

4. The ash formed when a volcano erupts is considered to be particulate matter.

5. The danger of radon can be minimized by good household ventilation.

6. Ozone is a secondary pollutant.

7. Nitrous acid is a primary pollutant.

8. Carbon monoxide is responsible for the brown haze in smoggy cities.

Short Answer

1. The EPA says that carbon monoxide levels should be maintained below 35 ppm for any one hour period. How many liters of carbon monoxide could legally be present in a 2000 liter sample at that concentration?

2. What is the concentration of particulate matter in $\mu g/m^3$ in a sample found to contain 0.0028 grams in 80.0 m^3 of dry air?

3. What is a) the percent by volume and b) the concentration in parts per million of a 7.5 liter air sample found to contain 0.015 milliliters of nitrogen dioxide?

4. Write proper products for the following combinations as discussed in the text and this guide and balance the equations.

a) $C + O_2$ --> (sufficient oxygen)

b) $C + O_2$ --> (insufficient oxygen)

c) $S + O_2$ -->   d) $SO_2 + O_2$ -->

e) $SO_2 + H_2O$ -->   f) $SO_3 + H_2O$ -->

g) $NO_2 + H_2O$ --> (two acids)

h) Cl_2F_2C --> (sunlight present)

i) $N_2 + O_2$ --> (high temperature)

j) $Cl + O_3$ -->

k) NO_2 --> (sunlight)

l) $O + O_2$ -->

5. ^{222}Rn is an alpha emitter with a half-life of 30 seconds.   a)   What is the product of the alpha emission?   b)   How much of a 1.00 gram sample of this isotope would remain undecayed in 2.0 minutes?

ANSWERS TO SELF-TEST

Fill in the Blanks

1. nitrogen; oxygen   2. mercury and beryllium
3. carbon monoxide; hydrocarbons
4. sulfur oxides; nitrogen oxides
5. ultraviolet 6. Radon   7. ozone; PANs
8. chlorofluorocarbons   9. 5.6   10. scrubbers

True or False

1. False--hemoglobin attachment   2. False--lung cancer
3. True   4. True   5. True   6. True
7. False--formed from H_2O and NO_2 in the air.
8. False--colorless gas, browns usually due to nitrogen containing materials or particulate matter

Short Answer

1. 0.070 L_3 (2000 x 35/1,000,000); 70 mL (0.070 x 1000)
2. 35 $\mu g/m^3$ (0.0028 x 1,000,000/80)

3. a) 0.00020 % ([0.015/1,000]/7.5 x 100)

   b) 2.0 ppm ([0.015/1,000]/7.5 x 1,000,000)

4. a) $C + O_2 \longrightarrow CO_2$

   b) $2C + O_2 \longrightarrow 2CO$

   c) $S + O_2 \longrightarrow SO_2$

   d) $2SO_2 + O_2 \longrightarrow 2SO_3$

   e) $SO_2 + H_2O \longrightarrow H_2SO_3$

   f) $SO_3 + H_2O \longrightarrow H_2SO_4$

   g) $2NO_2 + H_2O \longrightarrow HNO_2 + HNO_3$

   h) $Cl_2F_2C \xrightarrow{UV} ClF_2C + Cl$

   i) $N_2 + O_2 \longrightarrow 2NO$

   j) $Cl + O_3 \longrightarrow ClO + O_2$

   k) $NO_2 \longrightarrow NO + O$

   l) $O + O_2 \longrightarrow O_3$

5. a) $^{222}_{86}Rn \longrightarrow {}^{4}_{2}He + {}^{218}_{84}Po$

   b) 0.0625 g left (1.00 x 1/2 x 1/2 x 1/2 x 1/2)

122

12

ORGANIC CHEMISTRY

CHAPTER OVERVIEW

This chapter focuses on the chemistry of carbon compounds. Structural formulas for compounds and their isomers are written and the distinguishing characteristics of different saturated and unsaturated hydrocarbons are presented. Simple naming rules and some typical reactions are given. Major functional groups and examples of the name and use of compounds in each group are provided. Since different instructors emphasize different specific compounds and may prefer either IUPAC or common names, this study guide will only include a few particularly common examples.

TOPIC SUMMARIES AND EXAMPLES

Organic Compounds Defined

Organic compounds were originally thought to be only synthesized by living organisms, while **inorganic** compounds had nonliving origins. Synthesis of urea from an inorganic compound disproved the theory of **vitalism**. **Organic** compounds are now classified as compounds which contain carbon and various other elements, therefore, **organic chemistry** is considered to be the chemistry of carbon and its compounds.

The Amazing Carbon Atom and the Bonding Between Carbon Atoms

Carbon forms millions of compounds and is unique in that it is the only element capable of bonding to itself to an appreciable extent. The simplest organic molecule is methane, CH_4, in which carbon is bonded to the four hydrogen atoms in a **tetrahedral** shape. Carbon makes four covalent bonds (shares eight electrons) in covalent compounds and can bond with up to four other carbons in single, double or triple bonds.

Structural Formulas of Organic Compounds

The **molecular** formula of a compound tells the actual number of atoms of each element which comprise the compound. It may be possible to arrange the atoms in multiple ways. A **structural** formula of a compound shows the actual way in which the atoms are bonded to each other. This bonding can be shown in electron dot, dash or condensed notation.

Example 12.1

Propane (C_3H_8) may be shown as:

$$H : \overset{\overset{\displaystyle H}{\cdot\cdot}}{\underset{\overset{\displaystyle H}{\cdot\cdot}}{C}} : \overset{\overset{\displaystyle H}{\cdot\cdot}}{\underset{\overset{\displaystyle H}{\cdot\cdot}}{C}} : \overset{\overset{\displaystyle H}{\cdot\cdot}}{\underset{\overset{\displaystyle H}{\cdot\cdot}}{C}} : H \quad \text{or} \quad H - \overset{\overset{\displaystyle H}{|}}{\underset{\underset{\displaystyle H}{|}}{C}} - \overset{\overset{\displaystyle H}{|}}{\underset{\underset{\displaystyle H}{|}}{C}} - \overset{\overset{\displaystyle H}{|}}{\underset{\underset{\displaystyle H}{|}}{C}} - H \quad \text{or} \quad CH_3 - CH_2 - CH_3$$

(condensed)

The Classification of Organic Compounds and the R-Group

Organic compounds can be classified into groups with certain similar properties. To focus on the **functional group** which gives a particular compound unique properties, an **R** is often used to stand for the rest of the molecule, usually unreactant **alkyl** groups.

The Alkanes

The alkanes are the simplest organic compounds and contain only hydrogen and carbon with a general formula of $C_{2n}H_{2n+2}$. They are **saturated** compounds, which means that all bonds are single bonds. These compounds are fairly nopolar because of the similarity of electronegativity of carbon (2.5) and hydrogen (2.1). The alkanes form a **homologous series** in which the next group member has a structure which differs from the previous one by a constant increment, in this case a CH_2. Study Table 12-1 in the text for the listing of the family members. The table also provides the prefixes which reflect the number of carbons in the compound for the "ane" ending. Note that the alkyl group mentioned above is derived from an alkane from which a hydrogen is dropped so it can bond to another carbon. For example, ethane, CH_3CH_3, gives rise to an ethyl, CH_3CH_2-, R group.

All alkanes with four or more carbons will have **isomers**. These have the same molecular formula, but different structural formulas. The number of isomers increases dramatically with the number of carbons (see Table 12-2 in the text). Isomers can be either **straight** chain or **branched** chain. They are only different if the bonding sequence is actually different, not if they just look different when written on the paper.

124

Example 12.2

Are the following compounds isomers?

$$
\begin{array}{ccc}
& CH_3 & \\
& | & \\
CH_3-CH_2-CH & \quad \text{and} \quad & CH_3-C-CH_3 \\
& | & \\
& CH_3 &
\end{array}
\qquad
\begin{array}{c}
CH_3 \\
| \\
CH_3-C-CH_3 \\
| \\
CH_3
\end{array}
$$

Solution

Each has the formula C_5H_{12} and the carbons are differently attached to each other. These are isomers.

The Alkenes and Alkynes

The **alkenes** have double bonds, while the **alkynes** have triple bonds, so that both are considered to be unsaturated compounds. Their names relect their bonding. For example, **propyne** would have three carbons and a triple bond and **propene** would have three carbons and a double bond.

The Cyclic Hydrocarbons

Cyclic hydrocarbons are closed structures, often shown as geometric shapes with the actual carbon and hydrogen atoms omitted. They have a prefix of **cyclo** in addition to that indicating the number of carbons. They may be cycloalkanes, alkenes or alkynes.

Example 12.3

Cyclobutane has four carbons. It may be written as:

$$
\begin{array}{cc}
\begin{array}{ccc}
H & & H \\
| & & | \\
H-C & - & C-H \\
| & & | \\
H-C & - & C-H \\
| & & | \\
H & & H
\end{array}
& \text{or} \quad \square
\end{array}
$$

Aromatic Hydrocarbons: They Don't All Smell Sweet

Aromatic compounds are closed ring structures whose dot structures make it appear as though they have alternating double and single bonds. Actually, all of the bonds in the compound prove to be equivalent. Aromatic compounds are ordinarily shown as a closed geometric figure, with a circle in the center, to show bond equivalence. The parent of the family of aromatic compounds is benzene, C_6H_6:

Reactions of Hydrocarbons

The alkanes tend to be quite stable. Their most notable reaction is combustion with oxygen producing carbon dioxide and water. In fact, the primary use of alkanes has been as fuels. All other hydrocarbons, and, in fact, essentially all organic molecules are combustible.

Example 12.4

The combustion reaction of octane, C_8H_{18} is:

$$2C_8H_{18} + 25O_2 \longrightarrow 16CO_2 + 18H_2O$$

Alkenes and akynes are quite reactive at their multiple bonds. Each undergoes **addition** reactions with hydrogen to form an alkane in a process called **hydrogenation**. Cyclo-alkanes will react with hydrogen to open the ring and create the straight chain hydrocarbon.

Example 12.5

Propene, propyne and cyclopropane can all be hydrogenated to yield propane.

$$CH_3-CH=CH_2 + H_2 \longrightarrow CH_3-CH_2-CH_3$$

$$CH_3-C\equiv CH + 2 H_2 \longrightarrow CH_3-CH_2-CH_3$$

$$\triangle + H_2 \longrightarrow CH_3-CH_2-CH_3$$

Alcohols: The -OH Functional Group

The presence of the -OH group causes alcohols to be polar and to be able to undergo hydrogen bonding. The alcohols with a low number of carbons are totally miscible in water because of this factor. Alcohols with a higher number of carbons have a larger percentage of nonpolarity, and no longer dissolve in water, but instead become quite soluble in nonpolar solvents. Important alcohols include: methyl alcohol (methanol), also known as wood alcohol; ethyl alcohol (ethanol), the drinking alcohol; isopropyl alcohol (2-propanol), which is rubbing alcohol; and glycerol (1,2,3-propanetriol), in which the triol indicates that there are three -OH groups.

Ethers: The R-O-R' Compounds

In **symmetrical ethers**, the R group is the same as the R' group. In **asymmetrical ethers**, the two groups are different. Ethers are incapable of hydrogen bonding and tend to be quite volatile. Ethers can be made from alcohol in the presence of sulfuric acid which causes dehydration (removal of a water molecule).

Example 12.6

$$CH_3-CH_2-OH + HO-CH_2-CH_3 \xrightarrow{H_2SO_4} CH_3-CH_2-O-CH_2-CH_3 + H_2O$$

ethanol   ethanol   diethyl ether
                                          (symmetrical)

Diethyl ether is used as an anaesthetic and is the most widely known of the ethers.

Aldehydes: The $-\overset{\overset{\displaystyle O}{\|}}{C}H$ **Functional Group**

Aldehydes often have pleasant tastes and odors. They can be readily reduced with hydrogen to yield alcohols. They also are susceptible to oxidation which produces carboxylic acids.

Example 12.7

$$CH_3-\overset{\overset{\displaystyle O}{\|}}{C}H + H_2 \xrightarrow{Pt} CH_3-CH_2-OH \quad \text{(reduction)}$$

ethanal   ethanol

$$CH_3-\overset{\overset{\displaystyle O}{\|}}{C}H + O_2 \xrightarrow{\text{oxidizing agent}} CH_3-\overset{\overset{\displaystyle O}{\|}}{C}-OH \quad \text{(oxidation)}$$

ethanal   ethanoic (acetic) acid

Foul smelling formaldehyde, used for preservation of biological specimens, is the best known of the aldehydes.

Ketones: The $R-\overset{\overset{\displaystyle O}{\|}}{C}-R$ **Compounds**

The $-\overset{\overset{\displaystyle O}{\|}}{C}-$ group is called the **carbonyl or keto** group. Although these compounds have a similar structure to that of the aldehydes, the extra R group bonded to the carbonyl carbon makes them very difficult to oxidize. Acetone (propanone), used in nail polish remover, is the best known of the ketones.

Carboxylic Acids: **The Organic Acids With That** $-\overset{\overset{\displaystyle O}{\|}}{C}-OH$ **Group**

Carboxylic acids function as weak acids:

$$R-\overset{\overset{\displaystyle O}{\|}}{C}-OH + H_2O \longrightarrow R-\overset{\overset{\displaystyle O}{\|}}{C}-O^- + H_3O^+$$

Acetic (ethanoic) acid, found in vinegar, and formic (methanoic) acid, found in ant bites, are the most commonly encountered carboxylic acids.

Esters: The R-$\overset{\overset{\text{O}}{\|}}{\text{C}}$-O-R' Compounds

Esters often have pleasant smells and are often used as flavoring agents. They are formed by the **esterification** reaction of a carboxylic acid with an alcohol. Water is a byproduct of the reaction.

Example 12.8

$$CH_3-\overset{\overset{\text{O}}{\|}}{C}-OH \quad + \quad HOCH_3 \quad ---> \quad CH_3-\overset{\overset{\text{O}}{\|}}{C}-O-CH_3 \quad + \quad H_2O$$
ethanoic acid   ethanol   a methyl ester

Amines: The -N- Compounds

Amines are organic derivatives of ammonia and tend to have pungent odors. The amines are subdivied into three categories:

1) Primary amines, which are R-NH_2
2) Secondary amines, which are R-$\overset{}{N}H$-R'
3) Tertiary amines, which are R-$\underset{\underset{R''}{|}}{N}$-R'

The amines act as bases in water in the same way as does ammonia discussed in Chapter 8.

$$RNH_2 \quad + \quad H_2O \quad <==> \quad RNH_3^+ \quad + \quad OH^-$$

Amides: The R-$\overset{\overset{\text{O}}{\|}}{\text{C}}$-N- Compounds

Amides take on enormous importance in the formation of proteins. Amides also can be classed as primary, secondary and tertiary, using the same criteria (number of R groups attached to the nitrogen) as that used for amines. Amides are formed in a manner analagous to that of esters. A carboxylic acid reacts with a primary or secondary amine to yield the amide and release water in the process.

Example 12.9

$$CH_3-\overset{\overset{\text{O}}{\|}}{C}-OH \quad + \quad H_2N-CH_3 \quad --> \quad CH_3-\overset{\overset{\text{O}}{\|}}{C}-\overset{\overset{\text{H}}{|}}{N}-CH_3 + H_2O$$
ethanoic   methylamine   a methylamide
acid   (secondary)

SELF-TEST

Fill in the Blanks

1. The reaction of an acid and an alcohol produces a/an
    _____ and _____.

2. Carbon always makes _____ covalent bonds in
    organic compounds.

3. _____ act as bases in water.

4. _____ hydrocarbons have only single bonds.

5. In the process of combustion, a hydrocarbon reacts
    with oxygen to produce _____ and _____.

6. _____ have the same molecular formula, but
    different structural formulas.

7. The reduction of an aldehyde yields a/an _____.

8. _____ amines have no hydrogens attached to the
    nitrogen.

9. The general formula of an alkane is _____.

10. Aldehydes can be oxidized to _____.

11. An alkyne has a/an _____ covalent bond.

12. Methanol, ethanol, propanol and butanol are the first
    members of the _____ series of the _____
    group.

True or False

1. Amides contain oxygen.

2. $CH_3-CH=CH_2$ is an alkane.

3. $CH_3-CH_2-O-CH_3$ and $CH_3-O-CH_2-CH_3$ are isomers.

4. Cyclopentene contains 10 hydrogens.

5. Aldehydes can be reduced to alcohols.

6. Ethers are made from carboxylic acids and alcohols.

7. Alkynes undergo addition reactions.

8. CH_3-O-CH_3 and CH_3-CH_2-OH are isomers.

9. Ethers are capable of hydrogen bonding.

10. Decane has ten carbon atoms.

11. CH_3-CH_2-CH_2-OH is drinking alcohol.

12. Vinegar contains an aldehyde.

13. All organic compounds contain carbon.

Short Answer

1. Write the following formula in condensed form.

$$H-\overset{\displaystyle H}{\underset{\displaystyle H}{C}}-\overset{\displaystyle H}{\underset{\displaystyle H}{C}}=\overset{\displaystyle H}{C}-\overset{\displaystyle H}{\underset{\displaystyle H}{C}}-\overset{\displaystyle H}{\underset{\displaystyle H}{C}}-O-H$$

2. Name the group to which each of the following organic compounds belongs.

   a) CH_3-CH_2-CH_2-$\overset{\displaystyle O}{\overset{\|}{C}}$-$CH_2$-$CH_3$ b) CH_3-CH_2-$\overset{\displaystyle H}{\overset{|}{N}}$-$CH_2$-$CH_3$

   c) CH_3-CH_2-CH_2-$\overset{\displaystyle O}{\overset{\|}{C}}$-OH   d) CH_3-C $\equiv$ C-CH_2-CH_3

   e) CH_3-CH_2-$\underset{\displaystyle OH}{\overset{|}{CH}}$-$CH_3$ f) CH_3-CH_2-$\underset{\displaystyle O}{\overset{\|}{C}}$-$\underset{\displaystyle CH_3}{\overset{|}{N}}$-$CH_3$

   g) CH_3-$\overset{\displaystyle O}{\overset{\|}{C}}$-$OCH_2$-$CH_2$-$CH_3$ h) CH_3-CH_2-O-CH_3

   i) CH_3-CH_2-CH_2-CH_2-CH_3 j)

3. Write balanced equations for combustion of the following compounds.

   a) CH_3-CH_2-C   CH   b) CH_3-CH_2-CH_3

   c)   d)

4. Write structural formulas of the two isomers which have the molecular formula C_3H_6O. Label each with a family group.

5. Complete the following reactions and label the functional group of the product formed.

a) $CH_3-CH_2-CH_2-\overset{\overset{\displaystyle O}{\|}}{C}-OH$ + $H_2N-CH_2-CH_3$ -->

b) $CH_3CH_2C\equiv CH$ + H_2 -->

c) $CH_3-CH_2-CH_2-\overset{\overset{\displaystyle O}{\|}}{C}H$ + oxdizing agent -->

d) $H\overset{\overset{\displaystyle O}{\|}}{C}-OH$ + $HO-CH_2-CH_3$ -->

e) $CH_3-CH_2-CH_2-OH$ + $HO-CH_2-CH_2-CH_3$ $\xrightarrow{H_2SO_4}$

f) $CH_3-CH_2-CH_2-\overset{\overset{\displaystyle O}{\|}}{C}-OH$ + H_2O -->

6. Write the formulas of the three isomers of C_3H_8O and tell what kind of a compound each is.

7. Write the formula for cyclobutene.

Multiple Choice

1. Which of the following is an aromatic compound?

a) 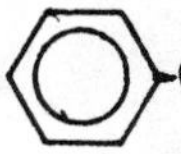 ⬡-CH_3 b) CH_3-O-CH_3 c) $CH_2=CH_2$ d) CH_4

2. Which compound exhibits hydrogen bonding?

a) CH_3-CH_3 b) $CH_3-\overset{\overset{\displaystyle O}{\|}}{C}-CH_3$ c) CH_3-OH d) all do

3. Which is an isomer of $CH_3-CH_2-CH_2-CH_3$?

a) $CH_3-\underset{|}{C}H_2$
    $CH_3-\underset{}{C}H_2$

b) $CH_3-CH_2-CH_2-CH_2-CH_3$ c) ▢

d) $CH_3-\underset{\underset{\displaystyle CH_3}{|}}{\overset{\overset{\displaystyle CH_3}{|}}{C}}H$

4. Which of the following compounds is unsaturated?

a) 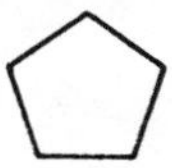 ⬠   b) CH_3-OH c) $CH_3-CH=CH_2$ d) CH_3-NH_2

131

5. Which of the following compounds is a secondary amine?

a) $CH_3-\overset{\overset{\displaystyle CH_3}{|}}{N}-CH_3$ b) $CH_3-CH_2-\overset{\overset{\displaystyle H}{|}}{N}-CH_3$ c) $CH_3-\overset{\overset{\displaystyle O}{\|}}{C}-\overset{\overset{\displaystyle H}{|}}{N}-CH_3$

d) CH_3-NH_2

6. Which is not a hydrocarbon?

a) 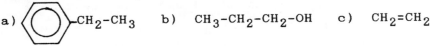$-CH_2-CH_3$ b) $CH_3-CH_2-CH_2-OH$ c) $CH_2=CH_2$

d) all are hydrocarbons

ANSWERS TO SELF-TEST

Fill in the Blanks

1. ester; water   2. four   3. amines   4. saturated
5. carbon dioxide; water   6. isomers   7. alcohol
8. tertiary   9. C_nH_{2n+2} 10. carboxylic acids
11. triple   12. homologous; alcohol

True or False

1. True--$\overset{\overset{\displaystyle O}{\|}}{C}$-N   2. False--alkene
3. False--same, just written differently
4. False--8 H   5. True   6. False--esters are
7. True   8. True--both C_2H_6O
9. False--no H attached to O   10. True
11. False--CH_3-CH_2-OH 12. False--acetic acid
13. True

Short Answer

1. $CH_3-CH=CH-CH_2-OH$

2. a) ketone   b) secondary amine
   c) carboxylic acid   d) alkyne   e) alcohol
   f) tertiary amide   g) ester
   h) asymmetrical ether   i) alkane   j) cycloalkane

3. a) $2CH_3-CH_2-C\equiv CH + 10O_2 \longrightarrow 8CO_2 + 6H_2O$

   b) $CH_3-CH_2-CH_3 + 5O_2 \longrightarrow 3CO_2 + 4H_2O$

   c) $2 \triangle + 9O_2 \longrightarrow 6CO_2 + 6H_2O$

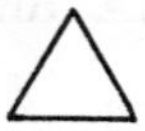

d) $2 \bigodot + 15O_2 \longrightarrow 12CO_2 + 6H_2O$

4.

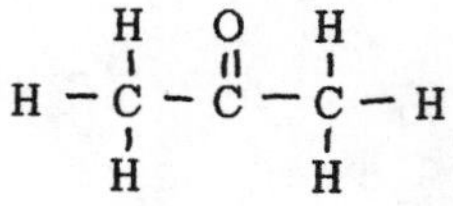

$$H - \underset{\underset{H}{|}}{\overset{\overset{H}{|}}{C}} - \underset{}{\overset{\overset{O}{\|}}{C}} - \underset{\underset{H}{|}}{\overset{\overset{H}{|}}{C}} - H \qquad \text{and} \qquad H - \underset{\underset{H}{|}}{\overset{\overset{H}{|}}{C}} - \underset{\underset{H}{|}}{\overset{\overset{H}{|}}{C}} - \underset{}{\overset{\overset{O}{\|}}{C}} - H$$

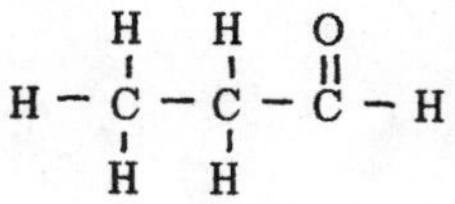

      ketone   aldehyde

5. a) $CH_3-CH_2-CH_2-\overset{\overset{O}{\|}}{C}-\overset{\overset{H}{|}}{N}-CH_2-CH_3$ (amide) $+ H_2O$

    b) $CH_3-CH_2-CH_2-CH_3$ (alkane)

    c) $CH_3-CH_2-CH_2-\overset{\overset{O}{\|}}{C}-OH$ (carboxylic acid)

    d) $H\overset{\overset{O}{\|}}{C}-OCH_2-CH_3$ (ester) $+ H_2O$

    e) $CH_3-CH_2-CH_2-O-CH_2-CH_2-CH_3$ (ether) $+ H_2O$

    f) $CH_3-CH_2-CH_2-\overset{\overset{O}{\|}}{C}-O^-$ (anion due to ionization) $+ H_3O^+$

6. $CH_3-\underset{\underset{OH}{|}}{CH}-CH_3$ (alcohol) $CH_3-O-CH_2-CH_3$ (ether)

    $CH_3CH_2CH_2-OH$ (alcohol)

7.

    □̲

Multiple Choice

  1. a   2. c   3. d   4. c   5. b   6. b

13

PLASTICS AND POLYMERS

CHAPTER OVERVIEW

The chapter discusses the history of polymer chemistry and the manner in which condensation and addition polymers form. The topics of elastomers and synthetic fibers are introduced. The difference between thermoplastic and thermosetting polymers is presented. Uses of polymers, environmental problems associated with polymers, and possible solutions for these problems are also discussed.

TOPIC SUMMARIES AND EXAMPLES

Introduction

Polymers are huge chains of organic molecules containing up to millions of atoms. Natural polymers include rubber, starch, cellulose, and proteins. Studies were begun to try to discover how the molecules were made and to determine if they and similar molecules could be synthesized.

A Short History of Big Molecules

This section discusses the process of **osmosis**, in which water moves from regions of lower solute concentration to regions of higher solute concentration in an attempt to equalize the concentration on both sides of a semipermeable membrane. The resulting **osmotic pressure** which develops enables the computation of molecular weight of macromolecules. The term **polymer** was used to refer to the many parts into which a macromolecule could be degraded.

The Nitrocellulose Story: Starting Off with a Bang

Nitrocellulose, in which natural cellulose from plant

fibers was treated with nitric and sulfuric acid, was the first **semisynthetic polymer**. It is considered to be only partially synthetic because the starting material, cellulose, is itself a polymer and the product, which found use as an explosive, was only a modification of an existing species.

Celluloid: The Discovery of Synthetic Plastic

Partially nitrated cellulose called **pyroxylin** was used to make **celluloid**, the first synthetic plastic. **Plastics** are substances which are capable of being molded into shape. Pyroxylin was also forced through tiny holes to produce **rayon** the first synthetic fiber.

Cellulose Acetate: The Genius of George Eastman

Cellulose was treated to accept acetate groups rather than nitrate groups. The resultant **cellulose acetate** was less flammable than the nitrocellulose and was used as a surface for photographic emulsions and for moving pictures.

The Discovery of Bakelite: The First Completely Synthetic Plastic

While previous work had modified a natural polymer, in this case, small units called **monomers** of **phenol** and **formaldehyde** were reacted to form an inert plastic called **Bakelite** which could be molded as it was formed. Bakelite is a **thermosetting plastic** which means that it cannot be remolded after being formed. It is also a **copolymer** since it is made of two different monomer units and a **condensation** polymer since water is eliminated in the polymerization process:

Example 13.1

The monomer units and polymerization to form Bakelite is:

phenol   formaldehyde

The reaction can take place at any position on the benzene ring, so a complex three dimensional structure is formed. See Figure 13-8 in the text.

The Rubber Revolution: The Elastomers

Natural rubber is a polymer of a single monomer called **isoprene**. It is an **addition** polymer in which polymerization occurs by adding to the double bonds. Instead of just hydrogen adding as was seen in the alkene reactions of Chapter 12, the whole next molecule adds to the bond. Its double bond can in turn be added to and the chain gets very long. No water is eliminated in addition polymers.

Example 13.2

The addition of isoprene units to form rubber is:

$$n \; \underset{\underset{\text{isoprene}}{}}{\overset{\text{H} \quad \text{H} \; \text{H}}{\underset{\text{H} \; \text{CH}_3 \; \text{H}}{\text{C=C-C=C}}}} \quad -> \quad \left(\overset{\text{H} \quad \text{H} \; \text{H}}{\underset{\text{H} \; \text{CH}_3 \; \text{H}}{\text{-C-C=C-C-}}} \right)_n$$

Rubber is called an **elastomer** because it is highly stretchable. Natural rubber gets sticky at high temperatures and brittle at cold temperatures. **Vulcanization,** in which sulfur is added and forms cross links by bonding between polymer chains, solves these temperature problems. **Neoprene,** in which a chlorine atom is substituted for the methyl group, is more resistant to certain chemicals than vulcanized rubber. A copolymer SBR which contains monomer units of styrene and butadiene is even more stable.

The Synthesis of Nylon: Another Condensation Polymer

Nylon was discovered in an attempt to imitate the peptide (amide) bond found in natural silk. Nylon is a copolymer of a dicarboxylic acid (adipic acid) and a diamine (ethylene diamine). Both molecules are capable of forming amide linkages at either end, and very long chains can be built. Water is eliminated in the process.

Example 13.3

The synthesis of nylon is:

$$n \; \underset{\underset{\text{ethylene diamine}}{}}{\overset{\text{H} \qquad\qquad \text{H}}{\underset{\text{H} \qquad\qquad \text{H}}{\text{N-CH}_2\text{-CH}_2\text{-N}}}} \quad + \quad n \; \underset{\text{adipic acid}}{\overset{\text{O} \qquad\qquad\qquad \text{O}}{\text{HO-C-CH}_2\text{CH}_2\text{CH}_2\text{-C-OH}}} \; --->$$

$$\left(\overset{\text{H} \qquad\quad \text{H} \; \text{O} \qquad\qquad\qquad \text{O}}{\text{-N-CH}_2\text{CH}_2\text{-N-C-CH}_2\text{CH}_2\text{CH}_2\text{-C-}} \right)_n \quad + \; n \; \text{H}_2\text{O}$$

Polyethylene: One of Our Most Important Polymers

This polymer is an addition polymer of ethylene (ethene).

Example 13.4

   The polymerization process is similar to that of rubber.

```
   H H            H H H H H H         / H H  \
   | |            | | | | | |        /  | |   \
 n C=C  --->   -C-C-C-C-C-C-  etc. or| -C-C-  |
   | |            | | | | | |        \  | |   /
   H H            H H H H H H         \ H H  /n
```

   ethylene

Polyethylene is a **thermoplastic** which means that it can be reheated and remolded repeatedly unlike the thermosetting plastics (such as Bakelite) which can only be molded when they are first formed. Polyethylene can be either **low density**, in which the polymer chains are rather randomly aligned, or **high density**, in which the chains are quite ordered. The first is much more flexible and finds use for soft plastic items. The high density form is employed when rigidity is of importance.

Polyvinylchloride (PVC): A Remarkable Polymer

   PVC is similar to polyethylene, except that one of the hydrogens is replaced by a chlorine. This is a tough plastic, widely used now for plumbing fixtures, since it will not rust or degrade as metals do.

The World of Synthetic Fibers

   In addition to nylon and rayon, many other synthetic fibers have been produced. Esterification (reaction with an acid) of -OH groups on cellulose produced rayon acetate. Various **polyester** fabrics have been produced, and have gained wide usage because of their durability. These condensation polymers form in a manner similar to that of nylon, except that a dialcohol is employed in place of a diamine. (Review the similarity of the process by which amides and esters are formed, presented in Chapter 12.) **Polyacrylonitrile** is discussed as being another addition polymer which forms in a similar manner to polyethylene.

Polymers and the Environment

   Problems associated with polymer production and usage are presented. Polymer production requires the use of a large amount of petroleum as starting materials and petroleum supplies are being rapidly depleted. The actual formation of polymers also requires a great deal of energy for synthesis. The desirable features of polymers, their resistance to chemicals, durability and inertness, are also detriments when the plastics are no longer needed. They do not readily degrade and have contributed to the mountains of waste the country generates. They are fire hazards, since they are composed of combustible carbon. In addition, disposal by incineration is dangerous because the fumes generated are often toxic. For example, PVC releases hydrogen chloride which, when dissolved in water, yields hydrochloric acid. Finally, many of the additives which impart specific properties to the polymers have proven to be toxic and can leach into the environment. Various recycling mechanisms and other methods are being used to try to minimize or eliminate some of these problems.

SELF-TEST

Fill in the Blanks

1. Polymers which have more than one kind of monomer unit are called _____.

2. In _____ polymers, water is eliminated as a byproduct.

3. Celluloid is a/an _____ polymer because it uses a natural polymer as a starting material for its production.

4. _____ forms the cross links in the vulcanization of rubber.

5. _____ are moldable polymers.

6. The _____ density polyethylene polymer is more flexible than the _____ density polymer.

7. _____ provides the raw materials for most plastic production.

8. _____ is the monomer unit of natural rubber.

9. Polyesters are made from _____ and _____.

10. _____ polymers are moldable only when they are first formed.

True or False

1. Nylon is an example of a copolymer.

2. Bakelite is a thermoplastic.

3. Rayon acetate is a synthetic fiber.

138

4. Polyvinyl chloride is an addition polymer.

5. Formaldehyde and phenol are the monomer units of Bakelite.

6. Plastics are inert to combustion and chemicals.

7. Molecular weight of polymers can be millions of g/mole.

8. Polyesters are addition polymers.

Short Answer

1. What monomeric substance is used to make this polymer (trade name--Teflon)?

```
    F F F F F F
    | | | | | |
   -C-C-C-C-C-C-etc.
    | | | | | |
    F F F F F F
```

2. Draw a section of the polymer (trade name Saran) made from the monomer unit:

```
   H Cl
   | |
   C=C
   | |
   H Cl
```

3. Draw the polymer whose repeating unit is:

```
        CH3 H
        |   |
        C===C
        |   |
        H   H
```

4. Write a a portion of the polymer chain which could be formed from the monomers:

```
   H          H                 O         O
   |          |                 ||        ||
   N-CH2CH2-N    and     HO-C-CH2-C-OH
   |          |
   H          H
```

5. What are the monomer units of this polymer?

```
    O       O              O       O
    ||      ||             ||      ||
  -C-CH2-C-O-CH2-CH2-O-C-CH2-C-O-CH2-CH2-   etc.
```

6. List five environmental problems associated with the production or use of polymers.

ANSWERS TO SELF-TEST

Fill in the Blanks

1. copolymers 2. condensation 3. semisynthetic
4. sulfur 5. plastics 6. low; high
7. petroleum 8. isoprene
9. diacids and dialcohols (groups at both ends)
10. thermosetting

True or False

1. True--adipic acid and ethylene diamine
2. False--thermosetting
3. False--derived from the natural cellulose polymer
4. True 5. True 6. False--burn readily
7. True 8. False--esterification releases H_2O

Short Answer

1.
```
F F
 |  |
 C=C
 |  |
F F
```

2.
```
     H  Cl H  Cl H  Cl H  Cl
     |  |  |  |  |  |  |  |
    -C--C--C--C--C--C--C--C-etc.
     |  |  |  |  |  |  |  |
     H  Cl H  Cl H  Cl H  Cl
```

3.
```
    CH3 H CH3 H CH3 H CH3 H
     |  | |  | |  | |  |
    -C----C-C----C-C----C-C----C-etc.
     |  | |  | |  | |  |
     H  H H  H H  H H  H
```

4.
```
     H        H O       O H       H O       O
     |        | ||      || |      | ||      ||
    -N-CH2-CH2-N-C-CH2-C-N-CH2-CH2-N-C-CH2-C-etc.
```

Water is released each time a N-C bond is made.

5.
```
    O        O
    ||       ||
 HOC-CH2-C-OH  and  HO-CH2-CH2-OH
```

6. 1) petroleum dependent 2) energy intensive
3) disposal problems 4) fire hazards
5) hazards of special additives

140

14

BIOCHEMISTRY: THE MOLECULES
OF LIFE

CHAPTER OVERVIEW

The chemistry and the function of the biological molecules, which include carbohydrates, proteins, nucleic acids and lipids, are presented. The structures of proteins in general and the special properties of enzymes are discussed. The nucleic acids RNA and DNA are polymers of nucleotides which have both similarities and differences. A survey of recombinant DNA, gene splicing and other new technologies completes the chapter.

TOPIC SUMMARIES AND EXAMPLES

Biochemistry is the chemistry of living organisms. It involves studying the composition and function of the different molecules which are needed to sustain life. All life forms are composed of one or more **cells** in which metabolism, which consists of all the life sustaining processes, takes place. **Metabolism** includes the breakdown of foods to use as building blocks and energy. It also includes the synthesis of complex molecules which are made from the pieces of nutrients. **Biomolecules** are the complex chemical compounds involved in all of these processes.

Carbohydrates

Carbohydrates, which are a primary energy source for the body's functions, are composed of sugars and starches. Both are basically **polyhydroxyaldehydes** or **polyhydroxyketones** or will release these compounds when reacted with water in a process called **hydrolysis**. Carbohydrates are classed according to the number of **simple sugar** or **monosaccharide** units they contain. A **dissaccharide** can be hydrolyzed to yield two simple sugars. A **polysaccharide** is a long polymer chain

141

which can be broken down into many simple sugar units.

The $C_6H_{12}O_6$ isomers **glucose** (blood sugar or dextrose), **fructose** (fruit sugar) and **galactose** are the most important monosaccharides. (See Figure 14-2 in the text for their structures.)

The common disaccharides are all $C_{12}H_{22}O_{11}$ isomers. They are condensation products of the simple sugars. (Note that adding the subscripts of two monosaccharides (listed above) and subtracting an H_2O gives the indicated molecular formula.) **Sucrose**, (table sugar) is the combination of glucose and fructose. **Lactose** (milk sugar) is composed of glucose and galactose. **Maltose** (malt sugar) has two glucose units.

All of the important polysaccharides are condensation polymers of glucose. **Starch** is digestible to humans. **Cellulose** is a building block of plants which, though not digestible to humans, provides dietary fiber. Humans and animals use **glycogen** to store excess glucose for future use. **Dextrin** is a component of glue and is responsible for many browning reactions in cooking. One gram of any carbohydrate, when metabolized by the body, releases 4 kcal.

Lipids

Lipids, which include fats and oils, provide another source of energy for the body. In the body, fats are used for insulation from temperature changes and to protect organs and tissue from injury. Lipids are insoluble in water, but soluble in nonpolar solvents such as benzene and carbon tetrachloride. Fats and oils are esters of **fatty acids** and **glycerol**, a trihydroxy alcohol. **Monoglycerides, diglycerides** and **triglycerides**, are esterified by one, two or three fatty acids, respectively. Natural fatty acids are synthesized by plants and animals two carbons at a time, so they always have an even number of carbons.

Example 14.1

The monoglyceride formed from glycerol and butyric acid is:

$$C_3H_7\overset{\overset{O}{\|}}{C}OH \quad + \quad \begin{matrix} HOCH_2 \\ | \\ HOCH \\ | \\ HOCH_2 \end{matrix} \quad ---> \quad C_3H_7\overset{\overset{O}{\|}}{C}-O-\begin{matrix} CH_2 \\ | \\ CH \\ | \\ HOCH_2 \end{matrix} \quad + \quad H_2O$$

butyric acid   glycerol   a monoglyceride

The acids in a fat may be the same or a mixture of various fatty acids. Hydrolysis (reaction with water in the presence of a catalyst) of a fat produces glycerol and the original fatty acids.

Fatty acids are **saturated**, if all of the carbon-carbon bonds are single bonds, or **unsaturated**, if at least one double bond is present. Animal fats are composed mainly of saturated fatty acids which are solid at room temperature. Vegetable fats, called **oils**, are liquid at room temperature because they are primarily composed of unsaturated fatty acids. **Hydrogenation** is the process by which hydrogen is added to some of the double bonds to make a firmer product as seen in vegetable **shortening**. Monounsaturated and poly-unsaturated fats are considered to be better for health than saturated fats. Each gram of a fat, when metabolized, provides 9 kcal (9 Calories) of energy, more than twice as much as the same mass of a carbohydrate.

Proteins

As a food, proteins provide energy (about 4.3 kcal/gram) and the building blocks to create more proteins. In the body, they are involved in numerous functions such as providing muscle tissue, transporting oxygen and regulating the activities of body processes. Proteins are condensation polymers of **amino acids** and may include other elements as well.

Amino Acids: The Building Blocks of Proteins

Amino acids have both a **carboxylic acid** group and an **amine** group. The general structure of an amino acid is:

```
    H R O
    | | ||
    N-C-C-OH
    | |
    H H
```

Each of the twenty amino acids found in proteins has a different **R** group. (See Table 14-4 in the text.) Some amino acids can be made in the body, but others, called the **essential amino acids**, must be ingested as part of the diet if the body is to be properly maintained.

Peptide Bonds

Peptide bonds or **peptide linkages** are the amide bonds which hold the amino acids together to make a **protein polymer**. A **dipeptide** is composed of two amino acids, a **tripeptide** of three amino acids and a **polypeptide** of many amino acids. Each time an amino acid is joined to the chain, a molecule of water is released.

Example 14.2

One dipeptide formed from glycine and alanine is:

$$
\begin{array}{c}
\text{H H O} \\
| \; | \; \| \\
\text{N-C-C-OH} \\
| \; | \\
\text{H H}
\end{array}
\; + \;
\begin{array}{c}
\text{H CH}_3 \text{ O} \\
| \; | \;\;\; \| \\
\text{N-C---C-OH} \\
| \; | \\
\text{H H}
\end{array}
\; --> \;
\begin{array}{c}
\text{H H O} \quad \text{CH}_3 \text{ O} \\
| \; | \; \| \quad\; | \;\;\; \| \\
\text{N-C-C-N-C---C-OH} \\
| \; | \quad\; | \; | \\
\text{H H} \quad \text{H H}
\end{array}
\; + \; H_2O
$$

glycine   alanine

In the other dipeptide which these amino acids could form, the positions would be reversed. Each end provides a site for continuation of the chain.

Amino Acid Sequence and the Function of Proteins

Table 14-6 in the text lists the ten main functions of proteins. Each protein has a characteristic sequence of amino acids which is essential for it to properly perform these tasks. Several genetic diseases are a result of signals which result in the formation of improper amino acid sequences and, consequently, the failure of a protein to perform as needed.

Protein Sequence: The Big Picture

The sequence of amino acids is called the **primary structure** of a protein. An average chain length is about 100 to 150 amino acids per molecule. Various methods have been used to determine the sequencing of specific proteins. One method involves reacting the protein with water under specific conditions. The hydrolysis reaction breaks the peptide bonds and releases the free amino acids.

The **secondary structure** of a protein is the arrangement of the protein polymer chain. Typical arrangements include a **helix**, which resembles a spiral staircase, a **pleated sheet** and a **random** or **disordered** configuration.

The **tertiary structure** of a protein is the way in which the secondary structure bends and folds in space. An example of secondary structure would be a coiled telephone cord. The tertiary structure would be the result of tying the spring in knots. The secondary structure is still apparent, but is warped. Hydrogen bonds are among the forces that maintain both the secondary and tertiary structures.

Quaternary structure is a characteristic of proteins which are comprised of more than one polymer chain. Each tertiary structure, called a **subunit,** is fitted into a pattern to form a complete protein unit.

Enzymes: A Very Special Group of Proteins

Catalysts are substances which increase the rate of a reaction, but are not themselves consumed. **Enzymes** are proteins which function as biological catalysts. In the body, each enzyme is responsible for catalyzing a specific reaction or class of reactions.

Enzyme Composition

Enzymes can be classified as **simple** or **conjugated** proteins. Simple enzymes have only protein material. Conjugated enzymes are composed of a protein and a nonprotein group called a **cofactor**. Cofactors are often metal ions or complex organic molecules. The latter are called **coenzymes**. A particular enzyme may require both a metal ion and a coenzyme to perform its tasks.

The Different Types of Enzymes

Enzymes are labeled according to the name of the substrate (reactant) upon which they act or for the type of reaction they catalyze. For example, a **lipase** acts on lipids, while a **hydrolase** causes hydrolysis to occur.

A Theory of How Enzymes Work

Enzymes will only work with certain substrates, a characteristic known as **substrate specificity**. The **lock and key** theory of enzyme activity proposes that an enzyme has a specifically shaped region called an **active site**. A substrate with the proper shape to fit that region can combine with the enzyme. The substrate reacts while attached to the enzyme and the product is released. The enzyme is then free to combine with another molecule of substrate and repeat the process.

Nucleic Acids: The Basis of Life

Chromosomes in the cell nucleus are composed of **genes** which carry all of the information to direct life processes from eye color to heartbeat. These genes contain specific **nucleic acids** which direct the synthesis of amino acids into proteins designed for certain purposes. The nucleic acids consist of **deoxyribonucleic acid (DNA)** and **ribonucleic acid (RNA)**. DNA stores and transfers the information and RNA directs and controls the actual protein synthesis which takes place in the **ribosomes**.

Components of Nucleic Acids

Nucleic acids are condensation polymers of **nucleotides**.

Each nucleotide is composed of: 1) a five carbon sugar; **ribose** in RNA and **deoxyribose** in DNA; 2) a **phosphate** group; and, 3) a **nitrogen containing base**. Nitrogen containing bases fall into the categories of **purines**, which are double-ring structures, and **pyrimidines**, which are single-ring structures. The purine bases are **adenine** and **guanine**. The pyrimidine bases are **cytosine**, **thymine** and **uracil**. Adenine, guanine and cytosine are components of both RNA and DNA. For the fourth base, DNA has thymine and RNA has uracil.

The Structure of the Nucleic Acids

In a nucleic acid, the polymer backbone is composed of alternating sugar and phosphate groups in which each nucleotide supplies a sugar/phosphate pair. The nitrogen bases are attached to the sugars and are not part of the backbone. DNA is composed of two **complementary** strands of nucleotide polymer arranged into a **double helix**. The helical arrangement is held in position by the hydrogen bonding of complementary bases. Adenine (A) and thymine (T) form one complementary pair and guanine (G) and cytosine (C) the other.

Example 14.3

We can write the complementary sequence for any nucleic acid segment by remembering the requisite pairings:

Segment	A	G	G	T	C	A	T	T	C
Complementary sequence	T	C	C	A	G	T	A	A	G

RNA is composed of single strands of nucleotide polymers. Although RNA contains a single strand, base pairings do occur. In RNA, uracil is the complementary base of the adenine.

The Replication of DNA

Damaged cells or cells needed for growth are created by **mitosis** (cell division). Each daughter cell needs to have its own DNA to carry information. In **replication**, a strand of DNA uncoils and separates into the two complementary strands. Each strand collects nucleotides and synthesizes a new complement. The helices reform with the result being two identical new DNA molecules, one for each daughter cell.

Recombinant DNA: The Process of Gene Splicing

Recombinant DNA techniques are also known as **gene splicing**. Usually, a ring of DNA, called a **plasmid** is taken from a bacteria cell. The DNA is cut open and a segment of DNA (a gene) from another source is inserted. As the bacteria divides and grows, the new gene is transferred to all daughter cells. The gene can direct the synthesis of a desired protein, such as insulin for use by diabetics.

Interferon, Recombinant DNA and the Future

One use of recombinant DNA techniques has been in the production of interferon. This compound has been used to treat viruses by triggering antiviral proteins which keep a virus from reproducing and attacking other cells. It may also have applications in the treatment of various cancers.

Human insulin, pituitary growth hormone, tPA for heart patients, interleukin-2 as an anticancer agent and tumor necrosis factor, another anticancer agent are compounds produced by recombinant DNA techniques. In addition, gene mapping to determine the areas on genes responsible for production of various proteins is being done. It is hypothesized that recombinant DNA techniques could be used to remove the "defective" genes responsible for some disease. They could then be replaced by the proper nucleotide sequence to ensure normal protein formation.

SELF-TEST

Fill in the Blanks

1. _____ are examples of vegetable fats which tend to be _____ at room temperature.

2. Amino acids which cannot be synthesized by the body are called _____ amino acids.

3. A triglyceride contains three _____ .

4. The _____ structure of a protein is how it is bent and folded in space.

5. Biochemical catalysts are called _____ .

6. _____ which carry specific information are found on the chromosomes.

7. Single ring nitrogen bases are called _____ .

8. The three common simple sugars are _____ , _____ and _____ . They are also called _____ .

9. Nucleic acids have _____ as their monomer units.

10. Ions or molecules needed for an enzyme to function

properly are called _____ .

11. _____ bonding is responsible for holding the double helix together.

12. The bonds which form when a protein is produced are called _____ bonds.

13. The _____ is the reactant upon which an enzyme works.

14. A polyunsaturated fat has many _____ bonds.

15. _____ is the process by which DNA is reproduced.

16. Proteins are manufactured in the cell in the _____ .

17. The two DNA strands in the double helix are _____ instead of identical.

18. Enzymatic specificity is explained by the _____ theory.

True and False

1. Purine bases form complementary pairs with pyrimidine bases.

2. Simple enzymes contain protein plus a metal ion.

3. Glycerol is a component of all fats.

4. Fructose is a six carbon sugar.

5. Uracil is a purine base.

6. An **ase** ending is characteristic of an enzyme.

7. Sugar releases more energy per gram than fat.

8. $CH_3(CH_2)_{16}-\overset{\overset{\displaystyle O}{\|}}{C}-OH$ is an unsaturated fatty acid.

9. $\begin{array}{c}\overset{\overset{\displaystyle O}{\|}}{C}-H\\ |\\ H-C-OH\\ |\\ H-C-OH\\ |\\ CH_2OH\end{array}$ is the formula for a simple carbohydrate.

10. 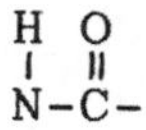 $\begin{smallmatrix} H & O \\ | & \| \\ N-C- \end{smallmatrix}$ is the peptide linkage.

11. Cellulose is an addition polymer.

12. Hydrolysis is the reaction of a substance with water.

13. Interferon interferes with virus reproduction.

Short Answer

1. How many kcal are provided by 75.0 g of carbohydrate?

2. How many kcal are provided by 75.0 g of fat?

3. Write the structural formula for glucose.

4. Write the formula of the diglyderide formed from
    $CH_3(CH_2)_{18}\overset{\displaystyle O}{\overset{\displaystyle \|}{C}}$-OH and glycerol.

5. Write the formula of the tripeptide formed from alanine, cysteine and glycine in the order written.

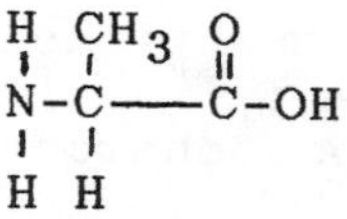

```
  H CH3 O              H  SH                 H H O
  | |   ||             | CH2 O               | | ||
  N-C——C-OH            N-C——C-OH             N-C-C-OH
  | |                  | |                   | |
  H H                  H H                   H H
  alanine              cysteine              glycine
```

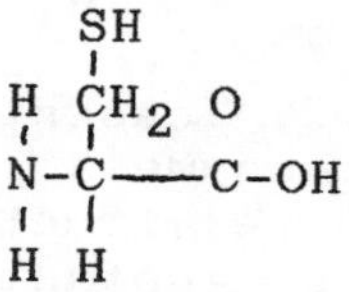

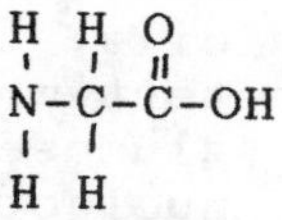

6. What is the complementary strand of this DNA segment?

    G C C T A A T T G C

7. What is the complementary strand of this RNA segment?

    G C A A U G C U C U

8. What are the amino acids formed from the hydrolysis of the compound below?

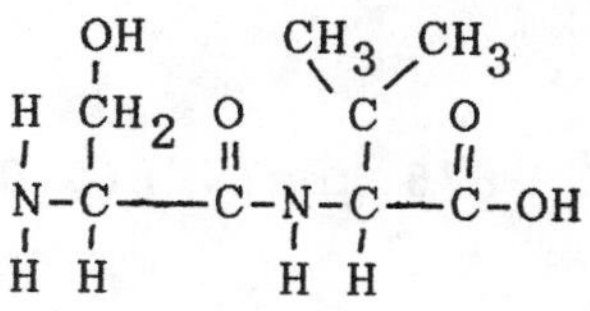

```
    OH          CH3  CH3
  H CH2 O        \  /
  | |   ||        C   O
  N-C——C-N-C——C-OH
  | |     | |
  H H     H H
```

9. What are the products when this fat reacts with water in the presence of a catalyst?

149

$$CH(CH_2)_5\text{-}\overset{\overset{\displaystyle H}{|}}{C}=\overset{\overset{\displaystyle H}{|}}{C}\text{-}(CH_2)_7\overset{\overset{\displaystyle O}{\|}}{C}\text{-}O\text{-}CH_2$$

$$CH_3(CH_2)_{14}\text{-}\overset{\overset{\displaystyle O}{\|}}{C}\text{-}O\text{-}CH$$

$$CH_3(CH_2)_3\text{-}(CH_2\underset{\underset{\displaystyle H}{|}}{C}=\underset{\underset{\displaystyle H}{|}}{C})_2\text{-}(CH_2)_7\text{-}\overset{\overset{\displaystyle O}{\|}}{C}\text{-}O\text{-}CH_2$$

10. What designations would be applied to the fat in question 9?

11. What are the hydrolysis products of each of the following substances?

   a) cellulose   b) maltose   c) lactose
   d) nucleic acids   e) proteins   f) sucrose
   g) fat   h) glycogen

12. List five differences between DNA and RNA.

ANSWERS TO SELF-TEST

Fill in the Blanks

1. oils; liquid   2. essential   3. fatty acids
4. tertiary   5. enzymes   6. genes   7. pyrimidines
8. glucose; galactose; fructose; monosaccharides
9. nucleotides   10. cofactors
11. hydrogen   12. peptide or amide
13. substrate   14. double   15. replication
16. ribosomes   17. complementary   18. lock and key

True or False

1. True   2. False--only protein   3. True
4. True   5. False--pyrimidine   6. True
7. False--fat 9 kcal/g compared to sugar 4 kcal/g
8. False--saturated   9. True   10. True
11. False--condensation; water is released
12. True   13. True

Short Answer

1. 300 kcal   (4.0 x 75.0)   2. 675 kcal   (75.0 x 9)

3.

$$\begin{array}{c} O \\ \parallel \\ C-H \\ | \\ H-C-OH \\ | \\ HO-C-H \\ | \\ H-C-OH \\ | \\ H-C-OH \\ | \\ CH_2OH \end{array}$$

4.

$$CH_3(CH_2)_{18}-\overset{\overset{\displaystyle O}{\parallel}}{C}\!-\!O\!-\!CH_2$$
$$\quad\quad\quad\quad\quad\quad O \;\; HO\text{-}CH$$
$$CH_3(CH_2)_{18}-\overset{\overset{\displaystyle \;\;}{\parallel}}{C}\!-\!-\!O\!-\!CH_2$$

5.

$$\begin{array}{cccccccc} & & & & SH & & & \\ & & & & | & & & \\ H & CH_3 & O & & CH_2 & O & & H & O \\ | & | & \parallel & & | & \parallel & & | & \parallel \\ N-C\!-\!-\!C-N-C\!-\!-\!-\!C-N-C-C-OH \\ | & | & & | & | & & | & | \\ H & H & & H & H & & H & H \end{array}$$

6. C G G A T T A A C G   7. C G U U A C G A G A

8.

$$\begin{array}{ccc} & OH & \\ & | & \\ H & CH_2 & O \\ | & | & \parallel \\ N-C\!-\!-\!C-OH \\ | & | \\ H & H \end{array} \qquad + $$

$$\begin{array}{ccc} CH_3 & CH_3 \\ \diagdown & \diagup \\ H & C & O \\ | & | & \parallel \\ N-C\!-\!-\!C-OH \\ | & | \\ H & H \end{array}$$

9. $HO\text{-}CH_2$ +

$$CH_3(CH_2)_5-\overset{\overset{\displaystyle H\;\;H}{|\;\;\;|}}{C}\!=\!\overset{}{C}-(CH_2)_7-\overset{\overset{\displaystyle O}{\parallel}}{C}-OH$$

$HO\text{-}CH$

$HO\text{-}CH_2$ + $CH_3(CH_2)_{14}-\overset{\overset{\displaystyle O}{\parallel}}{C}-OH$

             + $CH_3(CH_2)_3(CH_2\overset{}{C}\!=\!\overset{}{C})_2-(CH_2)_7-\overset{\overset{\displaystyle O}{\parallel}}{C}-OH$
                   $\quad\quad\quad\quad\quad\;\; H\;\; H$

10. polyunsaturated; triglyceride
11. a) glucose   b) glucose   c) glucose and galactose
     d) nucleotides   e) amino acids
     f) glucose and fructose
     g) fatty acids and glycerol   h) glucose
12. DNA--1) stores and directs information
          2) has deoxyribose   3) has thymine
          4) is double stranded
          5) is found in cell nuclei
     RNA--1) transfers messages and carries out synthesis
          2) has ribose   3) has uracil
          4) has a single strand
          5) moves from nucleus to carry out activities

151

15

FOOD CHEMISTRY

CHAPTER OVERVIEW

This chapter studies food chemistry beginning with the basic nutrients. The effect of cooking on the nutrient value of food is explored. The use of radiation to preserve food is mentioned in a scenario. The chemical basis of chocolate candy production is presented. Food additives used for both preservation and taste are described.

TOPIC SUMMARIES AND EXAMPLES

Introduction: You Are What You Eat

Proper nourishment is vital to the maintenance of a living system to supply both energy and the molecules needed to rebuild worn-out cells. The classes of nutrients are introduced. The use of gamma, cathode and X-rays is used to preserve food. The radiation damages living organisms such as molds and bacteria so that they cannot degrade the food. Careful control is needed since the radiation might initiate chemical reactions such as oxidation which causes rancidity of fats. Radiated food is monitored for **unique radiolytic products** which could cause adverse reactions if consumed.

Carbohydrates

Carbohydrates are the most abundant organic molecules on the earth and are also the body's preferred energy source. Foods which provide carbohydrates are grains, fruits and vegetables. In the absence of carbohydrates, the body uses its stored fats and proteins to power its life processes. Using fats can upset the balance of acids and bases in the body. Using up protein has serious adverse effects upon the body, because the body loses the building materials needed to replenish worn cells.

If more carbohydrates are consumed than the body needs for immediate energy, the remainder is stored as glycogen. Once glycogen reserves are full, any further excess is deposited as fat in the tissues. **Carbohydrate loading** is an attempt to fill the glycogen reserves before heavy physical activity is begun. The glycogen is much more quickly reconverted to glucose to be used for energy than are the stored fats.

Lipids

Lipids are found in oils, eggs, dairy products, meats and vegetables. They are also used as energy sources; however, it is recommended that no more than 30% of daily calories should come from fat sources. Since lipids are not soluble in water, they must first be **emulsified** and consequently can be complexed in the blood to form **lipoproteins**. Both **high density lipoproteins (HDL)** and **low density lipoproteins (LDL)** transport cholesterol. **Cholesterol** is a lipid which is not an ester of fatty acids and glycerol. It can be ingested but is also manufactured in the liver from fats. Cholesterol is essential for the formation of hormones and vitamin D, however, excesses have been implicated in heart disease and circulatory problems. Cholesterol will build up on interior blood vessel walls, restricting blood flow. HDLs are believed to be of benefit, since it appears that they can remove cholesterol from the walls and transport it to the liver for storage or excretion.

Proteins

Proteins are essential because they can be dismantled and their amino acids used to construct the proteins which the body needs for enzymes, muscles, hair, skin, etc. **Complete proteins** contain all eight of the essential amino acids and can be found in animal products and in soybeans. **Incomplete proteins** are lacking in one or more of the essential amino acids. Almost all vegetable protein is incomplete; mixtures must be eaten to supply all of the necessary amino acids.

Vitamins

Vitamins are primarily used by the body as **coenzymes** for the activity of specific enzymes. Only vitamin D can be formed in the body from cholesterol. The reaction is triggered by sunlight. Vitamins A, D, E and K are fat soluble and are stored in fatty tissues. Some problems can arise from taking too high a dosage of these vitamins because they will accumulate in the body. All of the B vitamins and vitamin C are water soluble so any excesses are excreted in the urine. Table 15-2 in the text lists dietary sources of the vitamins and tabulates the vitamin deficiency disease associated with each one.

Minerals and Water

Minerals include both metal and nonmetal ions. They are used in the body as coenzymes and also aid in regulating the fluid balance in the tissues. Calcium for bones, iodine for the thyroid gland and iron for the transport of oxygen are a few which are mentioned in the text. In addition, Table 15-3 lists other minerals and their usage. **Water** provides 60% of the body's weight. It provides a medium in which reactions occur, transports nutrients to tissues and removes wastes for disposal. Minerals and water maintain cell chemistry by subtle adjustments. Excesses of sodium can lead to high blood pressure because of the **osmosis** which results because of incorrect concentrations across the cell membrane.

Vitamins and Cooking

Cooking is used to kill organisms and to make chewing and digestion easier. It enhances flavor by releasing volatile materials which smell appetizing. If cooking water is discarded, boiling or steaming food leads to loss of any nutrients which leach into the water (especially water soluble vitamins and minerals). Heat destroys certain nutrients such as thiamine and hastens the oxidation of vitamin C. Cutting of food can shorten cooking times and cut down losses due to heat, but more of the soluble material then dissolves. Maximum nutritional value in cooked food can be maintained by short cooking times and by consuming all of the cooking broth.

The Chemistry of Chocolate

Production of chocolates begins with the fermentation and grinding of cocoa beans which contain protein and carbo-hydrates. The cocoa formed is sometimes treated with an alkaline solution to raise the pH and create Dutch chocolate. Lecithin is used to emulsify the nonpolar cocoa so that it can be dissolved in aqueous media such as milk. Soft centers are created by using an enzyme to hydrolyze sucrose to the more soluble glucose and fructose.

Which Milk is Best for Baby?

Human breast milk is designed for human babies. It is easily digested and contains lactose which aids absorption of minerals. Immunities can be transferred from the mother to the baby. Formulas attempt to mimic mother's milk by using cow's milk, to which many babies are allergic, or by using soy protein mixed with sucrose.

Food Additives

Food additives are used to improve the taste, appearence or shelf life of foods. The FDA maintains a **GRAS** list of **generally recognized as safe** additives which may be used without further testing. New products must gain approval before use in foods. **Preservatives** are needed because most food is consumed far from its place of origin and because few people shop daily. **Flavor enhancers** are used to intensify the taste of food. The most widely used flavor enhancer is **sodium chloride** (table salt), however, it has been linked to the incidence of high blood pressure and strokes. **Monosodium glutamate (MSG)** is another flavor enhancer, but some people experience adverse reactions to it.

The Inhibitor: Chemicals to Preserve Food Longer

Inhibitors retard mold and bacteria growth that spoil foods. **Nitrites** (NO_2^-) and **nitrates** (NO_3^-) have been used to maintain color and add spicy flavor as well as stop the growth of bacteria which cause botulism. Both are converted by hydrochloric acid in the stomach to **nitrous acid** (HNO_2) which then reacts with secondary amines to form **nitrosamines** which have been linked to stomach cancer.

Sulfur dioxide, though considered to be an air pollutant, has long been used as a disinfectant, a bleach and an antioxidant. It is used in wines, fruits and other substances to maintain color and prevent the growth of yeasts, molds and bacteria. Some people have problems with the sulfites formed by the dissolved sulfur dioxide gas.

BHA and **BHT** are used to prevent lipids from breaking down (becoming rancid) by acting as antioxidants. They have also been implicated in some physical abnormalities.

Sweeteners for Our Powerful Sweet Tooth

Natural sugars have a great deal of caloric value, but no other nutrients. Emphasis has been placed on finding sugar substitutes to give the sweet taste without unwanted calories. **Saccharin**, still in use, has what some people consider an unpleasant metallic taste and has been suspected of causing tumors. **Cyclamates**, which tasted better, have been banned in this country because of their link to cancer. **Aspartame**, made from the amino acids aspartic acid and phenylalanine, is now being widely used after intensive testing. It has problems because it is hydrolyzed in hot and acidic solutions and loses its sweetening ability.

SELF-TEST

Fill in the Blanks

1. HDL stands for _____.

2. The water soluble vitamins are _____ and _____.

3. _____ is considered to be an essential nutrient because it is a solvent for bodily processes.

4. Reserves of glucose are stored in the body as _____ and _____.

5. _____ proteins lack one or more essential amino acids.

6. _____ are a body's preferable energy source.

7. Cholesterol is useful because it is used in the synthesis of _____, _____ and _____.

8. _____ are essential to rebuild worn-out tissues.

9. _____ is a mineral linked to proper thyroid activity.

10. In liquid chocolate centers, _____ is converted by an enzyme into _____ and _____.

11. _____ is the most widely used flavor enhancer.

12. Nitrites have been implicated in the formation of carcinogens called _____.

True or False

1. Fat deposits provide more ready energy than does glycogen.

2. Vitamin C is the only vitamin which can be synthesized in the body.

3. HDLs are considered to be "good" lipids since they remove cholesterol from artery walls.

4. Antioxidants are used to keep fats from becoming rancid.

156

5. Vitamin K helps blood to clot.

6. Scurvy is caused by insufficient calcium.

7. Chocolate is a source of carbohydrates.

8. Mother's milk has more cholesterol than cow's milk.

9. Water soluble vitamins build up in the body.

10. Sulfur dioxide is a flavor enhancer.

11. BHT is an antioxidant.

12. Dutch chocolate is made by adding acid to cocoa.

ANSWERS TO SELF-TEST

Fill in the Blanks

1. high density lipoprotein
2. all B vitamins; vitamin C
3. water   4. glycogen; fat   5. incomplete
6. carbohydrates   7. hormones; vitamin D; bile salts
8. proteins   9. iodine
10. sucrose; fructose; glucose
11. sodium chloride   12. nitrosamines

True or False

1. False   2. False--vitamin D   3. True   4. True
5. True   6. False--vitamin C   7. True   8. True
9. False--fat soluble vitamins do
10. False--an antioxidant   11. True
12. False--alkaline or basic solution needed

16

AGRICULTURAL CHEMISTRY

CHAPTER OVERVIEW

This chapter deals with the process of food production. Nitrogen, potassium and phosphate fertilizers and the means of their application are presented. The reasons for trying to hybridize plants are discussed. The Green Revolution and subsequent increased dependence on insecticides and herbicides is stressed. Those methods are contrasted with those used by low-input organic farmers. The use of genetic engineering techniques for agriculture is presented.

TOPIC SUMMARIES AND EXAMPLES

Introduction

Many people are still undernourished in a world with a constantly growing population that will strain food supplies even more. This chapter looks at the fundamental techniques of agriculture which are being used to attempt to provide food with sufficient protein for these people.

How Do Plants Grow

Plants grow by the process of **photosynthesis**. This is an involved series of steps with an overall reaction of:

$$6CO_2(g) + 6H_2O(l) \longrightarrow C_6H_{12}O_6(s) + 6O_2(g)$$

This reaction requires both sunlight and chlorophyll to occur. You should recognize $C_6H_{12}O_6$ as being the formula of the simple monosaccharides of glucose, fructose and galactose. Much of the plant material occurs as polymers of glucose such as starches, cellulose and other complex carbohydrates. Note that this is the opposite of the net reaction that occurs when animals metabolize their food; therefore, the two types of organisms can benefit from each other.

Soil Fertilization and Nitrogen, Potassium and Phosphate Fertilizers

Soil needs to be enriched with the nutrients that growing plants deplete. The three soil nutrients that are needed in large quantities are nitrogen, potassium and phosphorus. Although air is 79% **nitrogen**, it can not be directly used by plants in the gaseous form. Instead it must be "fixed", which means that it must be reacted with another element to make a compound, usually **ammonia** (NH_3) or **nitrate** (NO_3^-) and **nitrite** (NO_2^-) salts or acids. Organisms such as algae and bacteria which live at the roots of plants called legumes (such as peas) can do this. These compounds can also be produced commercially from nitrogen in the air and are considered to be renewable. One process is the **Haber** process for making ammonia: $N_2(g) + 3H_2(g) \rightarrow 2NH_3(g)$ at high temperature and pressure. **Potassium** and **phosphorus** both must be mined and are considered to be nonrenewable nutrients.

10-6-4--What Does It All Mean

Labels on fertilizer bags and boxes have ratings with arrangement **N-K-P**, which stands for nitrogen, potassium and phosphorus. The numbers give the percent elemental nitrogen, the percent potassium, reported as K_2O, and the percent phosphorous, reported as P_2O_5. The 10-6-4 of the section heading is 10% nitrogen, 6% potassium and 4% phosphorus. Complete fertilizers contain all of these nutrients, while incomplete ones have only one or two. Nitrogen is used for leaf and stem growth. Phosphorus aids in root formation as well as root and flower development. Potassium helps increase plant resistance to disease. Lawns are treated with a high nitrogen fertilizer, because leaves are most important while fruits and vegetables require higher ratings of the other two nutrients. Problems have occurred in bodies of water due to runoff of these fertilizers which aid algae growth.

The Green Revolution: New Wonder Seeds

The **Green Revolution** has been a concerted effort to develop plant strains for maximum protein or other specified nutrient production. Crop vigor and the uniformity required by mechanical harvesting techniques are also goals. Worldwide production per acre has increased because of these new plant varieties and because of the intensified use of fertilizers and energy input. Genetic diversity decreases when only a few types of plants are grown and leaves huge fields subject to the same disease or pests. The National Seed bank has been established to maintain a source of different gene material for greater hybrid vigor.

Humankind Versus Pests: Protecting Our Food Supply

Pests are described as being any organism, such as rodents, certain insects and fungi, which damages crops. In the United States, more than one third of all food crops are lost to pests. It is very important to try to kill the pests without also harming beneficial animals such as bees, birds and earthworms.

DDT: The Start of the Modern Pesticide Era

DDT is a **chlorinated hydrocarbon** that was the first widely used **insecticide** (insect killer). It is readily and inexpensively produced.

The Effects of DDT

DDT worked very well as an insecticide and was widely used against the mosquitoes carrying malaria, however, resistant species began to develop. Additionally, the DDT is highly stable and is soluble in body fats and accumulates in tissues. It has proven harmful to birds, causing them to lay thin shelled eggs that break prematurely, so their offspring do not survive. DDT and other chlorohydrocarbons have been banned in the USA.

Other Ways of Killing Insects

Natural enemies, such as birds and other insects, can be used for control. Species specific viruses have been used against insects. Sterile males have been released and have mated with females instead of fertile males, thus stopping the birth of offspring. Hormones interfere with the reproductive cycle and sex attractants (**pheromones**) which lure insects into traps have also been employed. Insect resistant crops, such as corn with tight ears to exclude cornworms, have been developed to meet certain needs.

Herbicides: The Weed Killers

Herbicides are plant killers which been designed to attack specific plants. **Atrazine** works by blocking photosynthesis and **paraquat** kills weeds before the seedlings emerge. **Defoliants** such as **Agent Orange**, used in the Vietnam War, cause loss of leaves and consequent death of vegetation. Although plants perform better when they do not have to compete for nutrients, the herbicides do have problems. They have been implicated in miscarriages, nerve problems and other sicknesses. Dioxin, a byproduct in the production of the herbicide 2,4, 5-T has caused particular problems (see Chapter 1 on Times Beach crisis). Newer herbicides are being formulated to have fewer side effects, but so far have proved to be less discriminating between "good" and "bad" plants. It is a concern that herbicide resistant weeds are developing, consequently increasing the magnitude of the problem.

A Look at the Future of Agriculture

Natural pesticides and engineered organisms designed to combat specific weeds and pests are being studied. **Ice-minus** strain bacteria has already been developed to protect plants from frosts and is being used on fields. Corn and other grains are being manipulated in an attempt to have them "fix" their own nitrogen and lower fertilizer costs. **Bovine growth hormone** is being used to increase milk production. Additional work continues on safe genetic ways of increasing crop yields.

Food That's Organically Grown

Organically grown plants and animals are produced without use of inorganic fertilizers, herbicides or pesticides. Manures and composts of animal and plant wastes are used to provide nutrients. Natural means of repelling insects are used. Organic foods have the same nutrient content as inorganically grown food, but have no insecticide or herbicide residues.

Low-Input Agriculture

Low-input agriculture emphasizes organic methods, crop rotation and smaller plantings of more diversified crops. The advantage is lower production costs and few harmful residues, but increased personal labor is required.

SELF-TEST

Fill in the Blanks

1. To be used by plants, nitrogen must be _____. This is done naturally by algae and bacteria that live at the base of plants called _____.

2. A sexual attractant for insects is a/an _____.

3. Plant poisons are called _____.

4. The first widely used chlorinated hydrocarbon was _____.

5. 46-0-0 fertilizer is high in _____.

6. _____ is the process by which a plant produces food.

7. Agent Orange is a/an _____.

8. _____ and _____ are nonrenewable major nutrients.

9. _____ is being used to produce herbicide resistant crops.

10. Fertilizers derived from animal and plant sources are called _____ fertilizers.

True or False

1. The Green Revolution is heavily energy and chemical dependent.

2. Atrazine is a defoliant.

3. Ice-minus bacteria protect plants from frosts.

4. Dioxins are pesticides.

5. DDT has a long half-life in the environment.

6. Releasing sterile males is an organic way of controlling pests.

7. Bovine growth hormone is designed to put more meat on cows.

Short Answer

1. Write the equation for photosynthesis. What type of nutrient molecule is produced ?

2. Write the equation for the Haber process for synthesizing ammonia.

3. List four agricultural pests.

4. Why is DDT considered to be a special hazard.

5. Why is it said that plants and animals form a natural cycle?

6. What are the three main forms of fixed nitrogen?

7. The main source of hydrogen is petroleum. Why does the price of petroleum affect the cost of fertilizer?

8. A popular fertilizer is 8-8-8. Why?

ANSWERS TO SELF-TEST

Fill in the Blanks

1. fixed; legumes   2. pheromone   3. herbicides
4. DDT   5. nitrogen   6. photosynthesis
7. defoliant   8. potassium; phosphorus
9. genetic engineering   10. organic

True or False

1. True   2. False--blocks photosynthesis   3. True
4. False--byproducts of herbicide production   5. True
6. True   7. False--higher milk production

Short Answer

1. $6CO_2(g) + 6H_2O(l) \longrightarrow C_6H_{12}O_6(s) + 6O_2(g)$

    sunlight and chlorophyll are needed; monosaccharides (simple sugars) are formed; these are then often condensed into the various polysaccharides (such as starch and cellulose)

2. $N_2(g) + 3H_2(g) \longrightarrow 2NH_3(g)$

    (high temperature and pressure are needed)

3. rodents, insects, worms (larva of insects, not earthworms which are beneficial) and fungi
4. DDT is very stable and builds up in fatty tissues.
5. Plants produce sugar from water and carbon dioxide and release oxygen. Animals use oxygen and sugar to produce water and carbon dioxide so each makes use of the others products and, in turn, provides reactants for it.
6. ammonia (NH_3), nitrates (NO_3^-) and nitrites (NO_2^-)
7. The hydrogen is used in the Haber process of preparing ammonia, one of the main sources of nitrogen in fertilizers.
8. This gives a balance of the major nutrients to make sure that all growing needs are met.

17

PHARMACEUTICALS AND DRUGS

CHAPTER OVERVIEW

This chapter gives a history of drug use. It differentiates between chemical, generic and brand names. Antibiotics, antacids, birth control pills, stimulants, narcotics, tranquilizers and hallucinogens are detailed. Addiction and the dangers of mixing certain drugs or mixing drugs with certain foods or beverages are described.

TOPIC SUMMARIES AND EXAMPLES

Introduction

Drugs to combat infection and other conditions such as high blood pressure have greatly extended modern lives. Many ways of combatting high blood pressure are described. Enzyme inhibitors, beta blockers and diuretics are some of the methods mentioned.

Medicines and Drugs

Drugs are divided into four main categories: 1) fighting bacteria and other infections; 2) immunization agents; 3) affecting the operation of the circulatory system; and, 4) affecting the nervous system. Dollar amounts spent on legal drugs are listed. An estimate of the money expended on illegal drugs is also given. The most common over-the-counter (OTC) and prescription drug categories appear in Table 17-1 in the text.

Prescription Drugs

Each drug has a **chemical name** which is indicative of its components and structure. **Generic names** are shorter names used by the drug industry to describe the drug and **brand names** are the names by which the drugs are marketed and advertised. The book uses Valium as an example of the reason chemical names are not often employed. Brand names are more expensive and the companies argue that they are also prepared under more rigorous specifications. Table 17-2 in the text lists the ten most widely prescribed drugs in this country.

The Pain Relieving Drugs

Pain relieving drugs are called **analgesics**. Salicylic **acid** was the first widely used analgesic, in spite of its tendency to irritate the stomach. In addition, it was found to be an **antipyretic** (fever reducer). Modifications were made to improve it and the result was **acetylsalicylic acid** (generic name **aspirin**). Aspirin is used as an anti-inflammatory agent and an anticlotting agent in addition to its other properties. It has been proven to be extremely effective; however, people who are allergic to salicylates and people who have clotting problems should not take aspirin. Aspirin has also been implicated in Reye's syndrome which has caused death in some children who were being treated for high fevers.

Aspirin and Other Pain Relievers

More information is given on aspirin and it is noted that all brands of aspirin are equally effective. Aspirin works by suppressing the formation of **prostaglandins** which are involved in the inflammation process and in regulating blood pressure. **Acetominophen** is the leading aspirin substitute. It works well on pain and fever, but has no anti-inflammatory activity. It is recommended for children with high fevers and adults allergic to salicylates. **Ibuprofen** is the most recently approved nonprescription pain reliever. It works on pain, fever and inflammation and is especially effective against menstrual cramps. Preparations can be purchased which are combination drugs in which caffeine and antihistamines (to promote drowsiness) are also included.

The Antacids: Drugs for Stomach Upset

Overproduction of hydrochloric acid by the stomach causes irritation. Two main types of antacid are employed, one of which uses hydroxide ion and the other carbonate ion to neutralize acid. Combinations of the two are also employed.

Example 17.1

Old remedies for treating upset stomachs are baking soda ($NaHCO_3$) and milk of magnesia ($Mg(OH)_2$). They react with excess hydrochloric acid in the stomach as follows:

$$Mg(OH)_2 + 2 \ HCl \ ---> \ MgCl_2 + 2 \ H_2O$$

$$NaHCO_3 + HCl \ ---> \ NaCl + H_2O + CO_2$$

The Antibiotics: Life Saving Drugs

Chemotherapy in general is the use of a specific chemical to treat disease. **Antibiotics** are used to act against living organisms. The first use of antibiotics was synthetic **arsphenamine** to treat syphilis. **Sulfanilamide** was synthesized to fight various bacterial infections. It can do so because it resembles PABA which is needed by bacteria for folic acid synthesis. The drug attaches to an enzyme in the bacteria and prevents the production of folic acid. As a result, the bacteria die.

The first natural antibiotic was **penicillin** Synthetic pencillins were later developed with better antibiotic properties and less chance of allergic reactions. **Broad spectrum** antibiotics such as the tetracyclines were developed which could fight some viruses as well as bacteria. Resistant strains of bacteria have developed so efforts are being made to produce even more potent drugs.

Birth Control Pills and Synthetic Hormones

Hormones control various characteristics and functions in the body. Male hormones are called **androgens**, of which the most important is **testosterone** which controls the secondary sex characteristics such as facial hair. Women have two main categories of hormones: **estrogens**, of which **estradiol** and **estrone** are most important, and **gestagens**, of which **progesterone** is the most important. Estrogens control secondary sex characteristics and the menstrual cycle, while gestagens prepare the uterus for pregnancy. **Progesterone** was first used as an injectible **contraceptive** (means of preventing pregnancy) because it tricked the body into believing it was already pregnant. More recent developments are oral contraceptives which combine a menstrual regulator (estrogen) with a false pregnancy signal (gestagen). The tablets fall into the categories of **monophasic**, which give the same drug combination every day, **biphasic**, which change the concentration of two hormones throughout the month, and **triphasic**, which use a combination of three different drugs. Efforts are being made to produce a male birth control pill, but efforts so far have not been able to stop sperm production without also inducing unwanted side effects.

Steroids for Athletes: A Risky Business

Steroids have been developed for both **androgenic**

(masculinity) purposes and **anabolic** (muscle building) purposes. They have successfully been used to increase muscle mass, but have also caused side effects such as sterility and overagressiveness. Women can experience irreversible development of male sexual characteristics.

Cholesterol-Reducing Drugs

New drugs combined with a low fat diet have been found to significantly lower blood cholesterol levels. **HMG-CoA reductase** inhibitors reduce cholesterol by lowering LDL levels and have few side effects.

Stimulants

Caffeine is the most widely used stimulant and causes effects from alertness to hyperactivity. More potent stimulants are called "uppers", most notably **amphetamines** which were originally prescribed for weight loss. Amphetamines may be the most widely abused legal drug. **Cocaine**, a derivative of cocoa leaves, is the most widely abused illegal stimulant. Originally prescribed to cure alcohol addiction and for exhaustion, it was not thought to be addictive. Cocaine is now considered to be both psychologically and physiologically addictive. It causes an intense excitement or "high" when inhaled through the nose. **Crack** is a processed cocaine which is smoked and causes a more rapid and intense high, since it enters the blood stream more quickly. Addicted people can suffer mental problems such as hallucinations. Death can result from overstimulation. Work is being done to try to break the addictive cycle of this drug.

Depressants

Depressants, also called "downers", are used to calm anxieties or induce sleep. **Ethyl alcohol** (ethanol) is the most widely used depressant and can be lethal in high enough dosages. **Alcoholics** have an addictive reaction to ethanol, exhibiting both physiological and psychological dependencies. Work is progressing to find the source of the addiction and to eliminate it. **Barbiturates** can produce an effect ranging from mild sedation to deep sleep. People build drug tolerances to barbiturates and need higher and higher dosages to get the relaxing effect, often resulting in overdoses and death. Alcohol, combined with barbiturates, has a **synergistic effect**, which means that it enhances the extent of sedation and increases the possibility of a lethal overdose.

Narcotics

Narcotics are both depressants and analgesics. They have been traditionally considered to be addictive drugs. **Opium**, the first used narcotic, came directly from the opium poppy, whose leaves were chewed or smoked to get the desired effect. **Morphine**, the active ingredient in opium, was isolated and is still widely used as an extremely effective pain killer. **Codeine** is a derivative of morphine which is less addictive and can in fact be used at low concentrations in OTC drugs. **Heroin** is another derivative of morphine which has a more intense and addictive effect.

Tranquilizers

Tranquilizers are depressants designed to calm without inducing sleep. The most widely used drugs for relieving anxiety are the brand name drugs Librium, Valium and Miltown.

The Mind Altering Drugs

Mind altering drugs have various and unpredictable effects. LSD is an **hallucinogen** which produces extraordinary illusions. Additionally, hallucinations can recur without ingestion of additional dosages. PCP (angel dust) also has unpredictable side effects. **Marijuana**, with the active ingredient **tetrahydrocannibol (THC)**, produces milder effects. Some medicinal uses have been proposed for marijuana, which seems to aid people undergoing chemotherapy.

SELF-TEST

Fill in the Blanks

1. _____ are pain relievers.

2. _____ reduce pain and induce sleep.

3. _____ kill bacteria.

4. _____ is using chemicals to treat a disease.

5. A/an _____ effect is the intensification of one drug's effect by consuming another.

6. _____ steroids build muscle tissue and _____ steroids regulate male secondary sex characteristics..

7. _____ are fever reducers.

8. The first natural antibiotic was _____.

9. _____ are used for upset stomachs.

10. _____ work against bacteria by blocking the formation of folic acid.

11. _____ is the purpose of birth control pills which include compounds to _____ and to _____.

12. _____ is a more potent form of cocaine.

13. _____ calm without causing sleep.

14. _____is the pain reliever which also has anticlotting properties.

15. _____ is the active ingrediant in marijuana.

16. _____ are mind-altering drugs.

17. _____ hormones control women's secondary sex characteristics and _____ hormones prepare the body for pregnanacy.

18. _____, _____ and _____ are derivatives of the opium poppy.

Short Answer

1. Classify the following compounds according to the most appropriate category--analgesic, narcotic, antibiotic, hormone, stimulant, depressant, hallucinogen and tranquilizer.

   a) tetracycline   b) opium
   c) salicylic acid   d) barbiturates
   e) marijuana   f) testosterone
   g) caffeine   h) Valium
   i) cocaine   j) heroin
   k) estrol   l) PCP
   m) amphetamines   n) sulfa drugs
   o) ethanol   p) ibuprofen

2. $NaAl(OH)_2CO_3$ is present in a commonly sold antacid. Write its reaction with HCl.

3. What is a common property of aspirin and ibuprofen, but not of acetominophen?

4. Why is it recommended that certain drug combinations be avoided and that certain foods or beverages should not be consumed with certain drugs?

ANSWERS TO EXERCISES

Fill in the Blanks

1. analgesics 2. narcotic 3. antibiotics
4. chemotherapy 5. synergistic
6. anabolic; androgenic 7. antipyretics
8. penicillin 9. antacids 10. sulfanilamides
11. contraception; control menstrual cycle; create
    a false pregnancy
12. crack 13. tranquilizers 14. aspirin
15. THC 16. hallucinogens 17. estrogen; gestagen
18. heroin; morphine; codeine

Short Answer

1. a) antibiotic   b) narcotic
   c) analgesic   d) depressant
   e) hallucinogen   f) hormone
   g) stimulant   h) tranquilizer
   i) stimulant   j) narcotic
   k) hormone   l) hallucinogen
   m) stimulant   n) antibiotic
   o) depressant   p) analgesic

2. $NaAl(OH)_2CO_3 + 4\ HCl \longrightarrow NaCl + AlCl_3 + 3\ H_2O + CO_2$

3. both are anti-inflammatory

4. Certain combinations of drugs or food or beverages
   with drugs can decrease the effectiveness of the other
   drug. In addition, certain combinations can
   actually prove to be lethal.

18

THE CHEMISTRY OF HOME CARE AND PERSONAL PRODUCTS

CHAPTER OVERVIEW

This chapter discusses soaps and detergents and the means by which they work. Additives to enhance cleaning power are presented. Cosmetics and acne products and their ingredients are detailed. Fragrance products, antiperspirants, deodorants and their components are presented. Toothpastes, mouthwashes and hair care products complete the topic coverage.

TOPIC SUMMARIES AND EXAMPLES

Introduction

Soapwort or soap berries which produce lather were the first washing materials. Potassium and sodium carbonate, which make alkaline solutions in water, were then employed, with the latter still in use under the name of "washing soda". Soaps were originally made from fats (review chapter 14) and sodium hydroxide (lye), which could be obtained from wood ashes. The reaction is: fat + NaOH ---> glycerol + soap

The soap formed is of the general form: $CH_3(CH_2)_x\overset{O}{\overset{\|}{C}}-O^-Na^+$.

Potassium hydroxide is also used in soap making and produces a more liquid soap.

Soap Today

Commercial soaps are produced by hydrolysis of fats to release fatty acids and glycerol (see chapter 14). The fatty acids are then neutralized with NaOH or KOH, depending on the consistency of the product desired. **Abrasives**, which are gritty substances may be added for scouring power. Perfumes, deodorants and dyes are often added for aesthetic value.

How Soap Works

The carboxylic acid salt end of a soap molecule is ionic and is attracted to water (hydrophilic). The long hydrocarbon chain is nonpolar and is repulsed by water (hydrophobic). Many dirts and greases are nonpolar and are attracted to the nonpolar hydrocarbon portion of the soap. The dirts and greases get carried away in the water as the ionic end of the soap molecule clings to the water molecules. Soaps work well in "soft", alkaline water. In "hard" water, which contains calcium, magnesium and/or iron ions, the soap reacts with the ions in a double replacement reaction and precipitates as a scum. Acidic water causes the breakdown of a soap as is illustrated in the following reaction with hydrochloric acid:

$$CH_3(CH_2)_x \overset{O}{\overset{\|}{C}}-O^-Na^+ \ + \ HCl \ ---> \ NaCl \ + \ CH_3(CH_2)_x \overset{O}{\overset{\|}{C}}-OH$$

The fatty acids formed simply float as a grease on top of the water and cleaning ability is destroyed. Soap is a good cleaning agent in that is biodegradeable, nontoxic and is made from the renewable resource of fats and abundant sodium ion.

Synthetic Detergents: Syndets

Synthetic detergents (syndets) were developed to work in hard water. They work by the same general mechanism as does soap, with a polar end and a nonpolar end, but are formulated differently. **Sodium lauryl sulfate** was the first detergent, but was quite expensive. **Alkyl benzene sulfonates (ABS)** were less expensive, but had foaming problems. Since they were not biodegradeable, they would pass through water treatment plants unchanged and proved hazardous to the environment. **Linear alkyl sulfonates (LAS)** are biodegradeable. They have the general form:

$$CH_3(CH_2)_x-O-\overset{O}{\underset{O}{\overset{\|}{\underset{\|}{S}}}}-O^-Na^+$$

They are currently used in most detergents, however, there is some concern that they may cause formation of toxic phenol.

Surfactants and Builders: They Make Clothes Whiter and Brighter

Surfactants or **surface-active agents** include soap and detergents and include any substances which can suspend nonpolar materials in an aqueous medium. **Builders** increase cleaning ability, often by tying up hard metal ions to soften the water. **Phosphates** are excellent builders, but contribute to the problem of eutrophication by fostering the growth of

172

algae and other water plants (see Chapter 10). Their usage has been curtailed and phosphate detergents were actually banned in some communities. **Sodium perborate ($NaBO_2 \cdot H_2O_2$)** and **sodium metasilicate (Na_2SiO_3)** both work as builders. Both cause fewer environmental problems, but have been associated with skin irritation. Both react by releasing sodium hydroxide and raising the pH of the cleaning water. Sodium perborate also liberates **hydrogen peroxide (H_2O_2)** which has bleaching action.

Example 18.1

The reactions of the two builders above are:

$$NaBO_2 \cdot H_2O_2 + H_2O \longrightarrow NaOH + HBO_2 + H_2O_2$$

$$Na_2SiO_3 + 2H_2O \longrightarrow 2NaOH + H_2SiO_3$$

Anionic, Cationic, Nonionic and Amphoteric Surfactants

The previously discussed soaps and detergents were **anionic** surfactants, which means that they have negative charges at the nonhydrocarbon end. Surfactants can also be **cationic**, with a positive end, **nonionic**, with a neutral end, and **amphoteric**, which can act as either an acid or a base. Liquid laundry detergents usually are a combination of LAS and nonionic surfactants such as **alcohol ethoxylates**. The ethoxylates are esters of acids and are more soluble in cold than in hot water. Liquid dishwashing detergents contain the same basic ingredients as liquid laundry detergents, with more coloring and fragrance added. Automatic dishwashing detergents have very high pH's and are very caustic to the skin. **Fabric softeners** contain cationic surfactants. Cationic surfactants are usually quaternary ammonium salts, which are hydrocarbon derivatives of ammonia such as:

$$CH_3(CH_2)_{10}CH_2-\overset{\displaystyle CH_3}{\underset{\displaystyle CH_3}{N}}-CH_3 \quad Cl^-$$

These cationic surfactants are not good for cleaning since they stick to clothes. However, that sticking property is what makes them good fabric softeners. As they adhere they lubricate the fibers and prevent stiffness.

Bleaches and Brighteners

Surface electrons on clothes can use incident light to move to higher energy levels. Since the light is not reflected back, the clothes appear to be dull. Bleaches

173

brighten clothes by functioning as oxidizing agents, removing the surface electrons. Bleaches are reduced in the process. The oldest bleach used is **sodium hypochlorite**, in which the **hypochlorite** ion (ClO^-) is the active ingredient. It reacts as follows:

$$OCl^- + H_2O + 2\ e^- ---> Cl^- + 2\ OH^-$$

This also increases the cleansing activity of a soap since the hydroxide ion raises the pH as it is released. Hypochlorite bleaches have the unwanted effect of reacting with certain dyes and causing fading of fabrics. They also have the ability to cause fibers to break down, shortening the lifetime of clothing. **Hydantoin** and **cyanurate** bleaches are less potent, but avoid the fading and decomposition problem. **Hydrogen peroxide** acts as an oxidizing agent as it is reduced to water and oxygen gas. H_2O_2 works well for synthetics, but needs a highly alkaline solution and hotter water. Borax is often added to soaps or detergents to raise the pH to increase cleaning action. **Brighteners** contain **blancophors** which brighten the appearance of clothes by absorbing ultraviolet light and emitting visible light (fluorescence).

The Cosmetics of Yesterday

Perfumes, creams and powders have been used for years. Some, such as lead based powders, have had lethal results. **Cosmetics** are defined as products designed to promote attractiveness or to alter appearance. FDA approval is not required for their marketing.

Chemicals for the Skin: Lipsticks, Lotions and Creams

The skin is considered to be the largest organ of the body, protecting other tissues from infection and insulating them from environmental conditions. The skin also provides outlets for wastes and helps to regulate body temperature. Exposure to water, wind, sun and chemicals can cause dryness and irritation which are both unattractive and uncomfortable. Skin is usually protected or beautified by applying films to the surface. Lotions contain **emollients** to soften skin and **moisturizers** to prevent water loss. Petroleum jelly and mineral oil are nonpolar substances that serve both purposes. **Night creams** have the same constituents but are thickened with beeswax or paraffin to allow them to stay on for long periods of time. **Cleansing creams** contain the same ingredients as night cream. When they are wiped off, the nonpolar molecules carry with them not only dirt and grease, but also the residues of nonpolar makeups. **Lipsticks** contain more wax than oil to give them durability. **Antioxidants** may be added to all of the cosmetics to prevent the oils which they contain from

174

becoming rancid. All brands of cosmetics have the same basic ingredients. They vary only in perfume, coloring and specific proportions with which they are made.

Preparations for Acne

Acne is associated with the appearance of pimples and blackheads. It is caused by stress and hormonal changes and not by dirt, oils or grime, so cleaning and antibacterial preparations have little effect. **Benzoyl peroxide** has been successfully used to dry up pimples. **Retinoic acid** has worked very well, but has been implicated in various side effects. It has also been found to be successful at removing wrinkles in some cases.

Chemicals to Make Us Smell Better

Perfumes, colognes and **after-shave lotions** use ethyl alcohol as the solvent for odorous molecules. Perfumes contain ten to twenty-five percent odor bearing molecules, while colognes and after-shaves have only a few percent of these molecules. Components of perfumes are labeled by **notes**, where the notes refer to the volatility of the individual scent. The **top note** is most volatile and is smelled first, followed by the **middle note**. The **end note** is least volatile. See Table 18-1 in the text for examples of some compounds known for their aroma. **Fixatives** like **civetone** and **muscone** are used to mute some of the overly intense fragrances.

Chemicals for the Underarm Area

Sweat contains primarily water with trace amounts of oils. Bacteria can react with the oils to produce unpleasant odors. These can be controlled by simply washing the oil away. **Deodorants** cover up body odor with another scent. They also may contain antibacterial agents or other substances to cause the breakdown of oils. **Antiperspirants** contain an **astringent**, usually **aluminum chlorohydrate** which actually causes pores to contract to stop sweat from escaping, in addition to a deodorant.

Toothpastes and Mouthwash: Whiter Teeth and Cleaner Breath

Toothpastes contain detergents such as **sodium lauryl sulfate** and abrasives like calcium carbonate and titanium(IV) oxide. Although the sodium lauryl sulfate was considered too expensive as a laundry detergent, it is considered reasonably priced for the small amount of toothpaste used at any one time. The abrasives help remove plaque (bacteria growths) and surface food. Many toothpastes also have tin(II) fluoride (stannous fluoride) to counteract the formation of cavities.

The fluoride ions combine with the compound forming tooth enamel to form a harder, more impenetrable, surface. Toothpastes are always sweetened, flavored and sometimes colored for palatability. Thickeners are used as well. **Mouthwashes** may either be a deodorant, to cover up breath odor, or may contain a bactericide to kill odor-causing organisms. Bacteria have been implicated in tooth decay as well, and there is some evidence that mouthwashes can help fight the formation of cavities.

Chemicals for the Hair: Shampoos, Cream Rinses, Dyes, Permanent Waves and Straighteners

Hair is a protein which must be protected to maintain its properties and structure. **Shampoos** ordinarily are detergents, most often sodium lauryl sulfate, which remove oil and dirt. Colors, fragrances and thickeners are also shampoo components. Selenium or zinc compounds may be added for dandruff control. Buffers are used to maintain the pH in the range of five to eight which is compatible with the scalp and hair. **Baby shampoos** contain milder **amphoteric** surfactants. **Cream rinses** coat hair with additional protein. This acts as a lubricant to give a soft feel, but also attracts dirt and grease and necessitates more frequent washings. **Texturizers** are often glue-based to give body and to restore split ends. Hair color may be changed by bleaches, usually hydrogen peroxide, which removes all pigment, or by use of dyes. **Permanent dyes** penetrate the hair, while **temporary dyes** rinse out upon shampooing. Dye must be reapplied as the hair grows, because the new hair is the original hair color. The hair often needs additional care because of the stress of the dying process.

Hair proteins have a tertiary structure which is maintained in part by **disulfide (S-S)** bonds. Straighteners and permanent waves both work by adding a reducing agent to convert the disulfide bonds to SH groups. The hair is wrapped into the position desired and then reoxidized with hydrogen peroxide. The new disulfide bonds which form hold the hair into the desired position. The bleaching effect of the hydrogen peroxide often causes straightened or waved hair to be lighter than its natural color. Also, as in the case of dyes, as the hair grows, the original hair structure appears and the process must be repeated. **Mousses**, which are foams, and **hair sprays**, which are resins, are temporary methods of holding hair in desired positions.

SELF-TEST

Fill in the Blanks

1. _____ are used to increase the cleaning power of

soaps or detergents.

2. Soaps and detergents work best at a/an _____ pH.

3. Skin softeners are called _____.

4. Bleaches are _____ agents.

5. Lipsticks contain more _____ than do creams.

6. The carboxylic acid end of a detergent is _____ because it is attracted to water.

7. Antiperspirants contain _____ which deodorants do not have.

8. In permanent waves, _____ bonds are broken and reformed.

9. _____ have a higher percentage of odorous compounds than do _____ or _____.

10. Brighteners absorb _____ light and emit _____ light.

11. Cream rinses coat the hair follicles with _____.

12. The hardness of tooth enamel can be increased by the inclusion of _____ ions in toothpaste.

13. The _____ note of a perfume contains the most volatile compounds.

14. Soap doesn't work well in _____ or _____ water.

15. The builder which also bleaches is _____.

16. Mouth washes may contain _____ to kill odor causing organisms.

17. _____ tone down the scent of some perfumes.

18. The ideal pH range for shampoos is _____.

19. Fabric softeners contain _____ surfactants.
20. KOH produces a _____ soap than NaOH.

Short Answer

1. Give the name of the product or products associated

with the following substances.

a) sodium lauryl sulfate   b) petroleum jelly
c) abrasives   d) civetone
e) hydrogen peroxide   f) beeswax
g) sodium hypochlorite   h) alcohol ethoxylates
i) white glue   j) aluminum chorohydrate
k) cyanurate

2. What problem is associated with phosphate detergents?

3. How does a fabric softener work?

4. Why don't soaps work well in acidic or hard water?

ANSWERS TO SELF-TEST

Fill in the Blanks

1. builders   2. high or basic   3. emollients
4. oxidizing   5. wax   6. hydrophilic
7. astringents   8. disulfide
9. perfumes; colognes; after-shaves
10. ultraviolet; visible   11. protein
12. fluoride   13. top   14. hard; acidic
15. sodium perborate   16. bactericide   17. fixatives
18. 5-8   19. cationic   20. softer or more liquid

Short Answer

1. a) detergents, toothpastes, shampoos
    b) lipsticks, creams, lotions,
    c) toothpastes, soaps
    d) perfumes (as a fixative)
    e) fabric bleaches, hair bleaches, reestablishes
        disulfide bonds in straightening or waving hair
    f) night creams, lipsticks
    g) bleaches   h) liquid laundry detergents
    i) hair texturizers   j) antiperspirants
    k) bleaches
2. increases eutrophication in bodies of water
3. leaves a cationic coating on fabric to lubricate it
4. soaps precipitate in hard water and are converted to
    greasy fatty acids in acidic water

19

CHEMISTRY AND OUTER SPACE

CHAPTER OVERVIEW

The chapter deals with the "big bang" theory of how the universe was formed and looks at the chemical processes occurring in stars. Components of the sun and planets and discussion about the galaxies are also presented. Predictions of future events of the stars and universe are made. Space technology is seen as having contributed numerous consumer goods and methods for improving life.

TOPIC SUMMARIES AND EXAMPLES

The Universe

The **big bang** theory suggests that the universe began when a congregation of elementary particles exploded, spewing debris in all directions. Hydrogen gas coalesced to form the galaxies and at points condensed to stars which generate tremendous amounts of heat in fusion reactions. **Spectroscopes** are used to monitor the colors of light released and can determine the characteristics of the light emitting bodies. Our sun is composed of an interior **core**, where the fusion reactions occur, a **photosphere**, which is the bright layer from which we get the most light, and the thin outer layer, the **chromosphere**, which is characterized by flaming outbursts of gas. **Sunspots** are intense magnetic fields which migrate and have an eleven year cycle. Rotation of the sun is on a 27 earth-day cycle. The sun is marked by **prominences**, **flares**, and dark **coronal holes**. The last may be responsible for **solar wind** which causes brilliant **auroras** on earth.

The nine planets of the solar system are described. Our sun and solar system are characterized as being near the center of the **Milky Way** galaxy. Clouds of dust, called nebulae, may be the seeds of new stars or the remnants of old stars. Shapes of galaxies are mentioned as are **quasars** which emit enormous amounts of energy.

Stars are said to be in constant battle between the energy of the fusion reactions pushing mass **outward** and the gravitational attraction of the huge internal mass pulling matter **inward**. A star's life is marked by compression and expansion as the two forces compete. **Supernovas** and **black holes** are discussed as well. The immensity and complexity of the universe is stressed. The authors conclude the section with the view that radiowaves might be the best clue to the presence of other life forms in the universe.

The Energy of Nuclear Reactions

Nuclear reactions involve enormous changes in energy. According to Einstein's equation, $E = mc^2$, there is a relationship between energy and mass. In the equation, E is the energy in joules, m is the mass in kilograms and c is the speed of light, which is expressed as 3.0×10^8 m/s. If the change in energy of a reaction is known, the amount of mass which must have been converted to energy can be determined and vice versa.

Example 19.1

The heat of formation of H_2O (l) is -68.315 kcal/mol. What is the mass change in this process?

Solution

Since energy is released (shown by the negative sign) mass must decrease. First, the kcal must be changed into joules.

$$? \text{ joules} = -68.315 \text{ kcal} \times \frac{1000 \text{ cal}}{1 \text{ kcal}} \times \frac{4.184 \text{ J}}{1 \text{ cal}} = -2.858 \times 10^5 \text{ J}$$

$$m = \frac{E}{c^2} = \frac{-2.858 \times 10^5 \text{ J}}{(3.00 \times 10^8)^2} = 3.18 \times 10^{-12} \text{ kg}$$

This mass change is far to small to be detected by ordinary weighing devices.

Example 19.2

What is the energy in kJ released when one mole of helium-4 is made according to the fusion reaction below? The actual masses of the isotopes involved appear below the appropriate symbols.

$$2 \quad {}_{1}^{2}\text{H} \qquad ---> \qquad {}_{2}^{4}\text{He}$$

Actual mass   2.01410   4.00260

Solution

The change in mass in the total mass of the products minus the total mass of the reactants.

$$m = 4.00260 - 2(2.01410) = -0.0256 \text{ g}$$

$$\text{kg} = -0.0256 \text{ g} \times \frac{1 \text{ kg}}{1000 \text{ g}} = -0.0000256 \text{ kg} = -2.56 \times 10^{-5} \text{ kg}$$

$$E = mc^2 = (-2.56 \times 10^{-5} \text{ kg})(3.00 \times 10^8 \text{ m/s})^2$$

$$E = -2.30 \times 10^{12} \text{ J}$$

For any isotope, the isotope mass is **less** than the sum of the masses of the subatomic particles which compose it. This loss of mass is called the **mass defect**. That mass loss represents the **binding energy** released when the atom is formed. Conversely, binding energy also represents the energy that would be required to separate an atom into all of its individual pieces. The most stable elements have mass numbers between 30 and 63. They have the largest mass defects and, consequently, the highest binding energies.

Example 19.3

The actual mass of a fluorine-19 isotope is 18.9984032 g/mol. The masses of a proton, neutron and electron are, respectively, 1.00728, 1.00867 and 0.000549 g/mol. What is the mass defect?

Solution

The sum of the masses of the individual particles is:

9 protons	=	9(1.00728)	= 9.06552
10 neutrons	=	10(1.00867)	= 10.0867
9 electrons	=	9(0.00549)	= 0.04941

total mass of particles   = 19.2016

mass defect = actual mass - mass of subatomic particles

181

mass defect = 18.9984032 - 19.2016 = -0.2032 g

This mass defect may not look like a large number, but it represents an enormous amount of energy.

Example 19.4

What is the binding energy in the previous example?

Solution

The binding energy is $E = mc^2$. First, the mass must be converted to kilograms.

$$kg = -0.2032 \text{ g} \times \frac{1 \text{ kg}}{1000 \text{ g}} = -0.0002032 \text{ kg}$$

Now the energy equation can be used.

$$\text{binding energy} = -(0.0002032 \text{ kg})(3.00 \times 10^8 \text{ m/s})^2$$

$$\text{binding energy} = -1.83 \times 10^{13} \text{ J}$$

This is equivalent to the energy released by burning a thousand tons of coal.

Spin Offs from Space

The technology for space missions has had commercial and practical applications here on earth. A special insulation material was developed which helps oil stay warm and fluid enough to flow on the Alaska pipeline. Films to filter out infrared (heat) rays and insulate windows were developed. Vacuum drying of water damaged materials was perfected. An automated system for identifying microorganisms in body fluids has been used. A new alloy which can be bent repeatedly and still maintain its structural strength was developed. Smoke detectors which work on ionization or with photoelectric sensors have been designed. Liquid-cooled garments have been successfully used for burn victims. Chemists had used nuclear magnetic resonance (NMR) for the determination of the structure and characteristics of chemical compounds. Techniques used by the space program to filter immense quantities of data have been used with NMR to detect and pinpoint the location of diseased or damaged tissue in the body without the side effects of radiation. Storage of electrical energy has been made more inexpensive with **REDOX**, a system of membrane electrodes. REDOX works much like a car battery but, with no solids to fall away, recharging is more efficient and more complete. A method of automatic analysis of soil and rock samples has been devised. A new, durable, nontoxic, water based, inorganic paint was developed to protect the buildings belonging to NASA was then used to seal and protect the Statue of Liberty from future corrosion.

SELF-TEST

Fill in the Blanks

1. The reaction that the sun uses to generate energy is
   _____.

2. _____ are intense magnetic fields on the sun.

3. _____ may cause auroras on earth.

4. The _____ theory is a widely accepted view of the formation of the universe.

5. A star undergoes a constant battle between _____ pushing outward and _____ pulling inward.

6. The most stable elements have mass numbers between _____ and _____. These elements have the greatest _____ and _____.

Short Answer

1. What is the mass change in the following reaction? It generates 70.0 kcal of energy.

   $SiH_4(g) + 4 HCl(g) \longrightarrow SiCl_4(l) + 4 H_2O(g)$

2. Why do we say that mass in conserved in chemical reactions?

3. Calculate both the mass change and the energy in joules released in the following reactions. Actual masses of the species appear beneath them

   a) ^6_3Li $+$ ^2_1H $\longrightarrow$ $2 \, ^4_2He$

       6.01513   2.01410   4.00260   in g/mol

   b) $^{27}_{13}Al$ $+$ ^2_1H $\longrightarrow$ $^{25}_{12}Mg$ $+$ ^4_2He

     26.98153   2.01410   24.98584   4.00260

4. What can be said about the magnitude of energy changes in nuclear reactions compared to those in chemical reactions?

5. What are both the mass defect and the binding energy of the following atoms? The masses of a proton, a neutron, and an electron are, respectively, 1.00728, 1.00867 and 0.000549 g/mol.

a) $^{28}_{14}$Si   actual mass 27.97693 g/mol

b) $^{80}_{36}$Kr   actual mass 79.9164 g/mol

c) $^{238}_{92}$U   actual mass 238.0508 g/mol

ANSWERS TO SELF-TEST

Fill in the Blanks

1. fusion   2. sunspots   3. solar wind   4. big bang
5. energy; gravity
6. 30; 63; mass defect; binding energy

Short Answer

1. 3.25×10^{-12} kg   ($[70.0 \times 1000 \times 4.184]/[3.00 \times 10^8]^2$)
2. The mass losses or gains which occur are too small to be monitored with ordinary weighing devices.
3. a) -0.0240 g ($2[4.00260] - 6.01513 - 2.01410$)
      -2.16×10^{12} J ($[-0.0240/1000] \times [3.00 \times 10^8]^2$)

   b) -0.00719 g ($24.98584+4.00260-26.98153-2.01410$)
      -6.47×10^{11} J ($[-0.00719/1000] \times [3.00 \times 10^8]^2$)

4. The energy changes are enormously greater in nuclear reactions than in chemical reactions.

5. a) -0.25406 g; -2.29×10^{13} J

      $27.97693-(14[1.00728]+14[1.00867]+14[0.000549])$
      ($[-0.25406/1000] \times [3.00 \times 10^8]^2$)

   b) -0.7469 g; -6.72×10^{13} J

      $79.9164-(36[1.00728]+36[.000549]+44[1.00867])$
      ($[-0.07469/1000] \times [3.00 \times 10^8]^2$)

   c) -1.9353 g; -1.74×10^{14} J

      $238.0508-(92[1.00728]+92[0.000549]+146[1.00867])$
      ($[1.9353/1000] \times [3.00 \times 10^8]^2$)